INTRODUCTION TO
POLYMER

THE WYKEHAM SCIENCE SERIES
for schools and universities

General Editors:

Professor Sir Nevill Mott, F.R.S.
Cavendish Professor of Physics
University of Cambridge

G. R. Noakes
Formerly Senior Physics Master
Uppingham School

The aim of the Wykeham Science Series is to broaden the outlook of the advanced high school student and introduce the undergraduate to the present state of science as a university study. Each volume seeks to reinforce the link between school and university levels. The principal author is a university professor distinguished in the field assisted by an experienced sixth-form instructor.

INTRODUCTION TO
POLYMER SCIENCE

L. R. G. Treloar
Department of Polymer and Fibre Science,
University of Manchester
Institute of Science and Technology

WYKEHAM PUBLICATIONS (LONDON) LTD
LONDON AND WINCHESTER

SPRINGER-VERLAG NEW YORK INC.

1970

Sole Distributor for the United States, Canada and Mexico
SPRINGER-VERLAG NEW YORK INC./NEW YORK

Cover illustration – Fibrillar fracture of stretched crystalline rubber. This sample of raw rubber has been crystallized by stretching at room temperature, then cooled in liquid nitrogen before being fractured by striking with a hammer.

Title No. 8-88009

Library of Congress Catalog Card Number 70-116974

First published 1970 by Wykeham Publications (London) Ltd.

Printed in Great Britain by Taylor & Francis Ltd.
10–14 Macklin Street, London, WC2B 5NF

PREFACE

THIS book is intended primarily for sixth-form readers in schools, but should also meet the needs of students wishing to obtain a general knowledge of polymer science at a comparatively elementary level. Its writing has been carried out in close collaboration with Mr. W. F. Archenhold, Lecturer in the Department of Education at the University of Leeds, and formerly Head of the Physics Department, Huddersfield New College. His knowledge and experience in the field of education have been of great value, and I would like to express my appreciation of the advice and criticism which he has been able to offer during the preparation of the text.

My thanks are also due to Professor A. Keller for the photograph of polymer single crystals taken with the electron microscope, to Dr. I. H. Hall for the X-ray diffraction photographs of rubber and to Dr. J. Mann for the X-ray diffraction photographs of polypropylene. Finally, I have to acknowledge the work of Mr. G. B. Shipton, to whom I am indebted for the execution of the drawings.

The writer of a book of this kind, which endeavours to interest a wide range of readers, is faced with the problem of conveying in straightforward language an understanding of the broad principles underlying his subject, without undue reliance on the profusion of technical terms ('jargon' to the non-specialist) within which any branch of science tends to become encrusted. While it is clearly essential that new terms should be invented to express the new concepts of a developing science, too much dependence on specialized terms may be an embarrassment rather than an aid to understanding for those who have not yet learned the vocabulary. I have therefore tried to restrict the number of specialist terms to the minimum required for the purpose of expounding the subject and making contact with the accepted literature, and where such terms are introduced I have endeavoured as far as possible to explain them, and also to indicate their origin.

Manchester
January, 1970   L. R. G. TRELOAR

INTRODUCTION

THE main purpose of this book is to show how the physical properties of the principal types of polymers may be understood in terms of their basic molecular structure. It begins by examining the nature of the polymer molecule, i.e. its size and geometrical form, and continues by considering the way in which the particular features of the molecule determine whether a given polymer will be of the rubbery, glassy or crystalline type. The structure and physical properties of each of these classes of polymers (and also of the fibres, which are a special kind of crystalline polymer) are examined in some detail.

This presentation should enable the reader to understand not only the characteristic differences between the various kinds of polymeric materials of industrial importance, but also the broad differences in both structure and properties between polymeric and non-polymeric substances.

CONTENTS

CHAPTER 1
from rubbers to glasses

1. *The range of polymer materials*

POLYMERS are as old as man himself—indeed very much older—since in one form or another they are a basic constituent of every kind of living matter, whether plant or animal. It is only in the present century, however, that as a result of a wide range of scientific studies their existence as a coherent group has come to be recognized and understood. Following this recognition and this understanding, the possibility has arisen of actually producing polymers by means of suitable chemical reactions. Originally these synthetic products tended to be regarded as substitutes for existing natural polymers, such as rubber or silk, which were in short supply, but the more recent development of the polymer industry, dating roughly from the 1939–45 war, has led to the introduction of a vast range of entirely new compounds in the field of plastics, rubbers and fibres, many of which have properties different from those of any existing natural materials. The present-day study of polymers, though it includes the original natural polymers, tends to be dominated by these synthetic materials, since it is from the problems associated with the industrial development of such polymers that the main stimulus to scientific research has arisen.

In setting out to examine the scientific basis of polymers, we shall first have to consider what it is that all the various materials known as polymers have in common; what it is that constitutes a polymer. This common feature will be found to lie in the structure of the individual molecules of which the polymer is composed. We shall then take a broad look at the distinguishing features of the different classes of polymers, e.g. rubbers, fibres, etc., and will see how the outstanding physical properties of any one class of material are related to certain general features of their chemical or molecular constitution and to the way in which the molecules are arranged or held together to form a structure.

These basic concepts will be outlined in the present chapter, the object being to give the reader some feeling for the subject and some idea of its range and of the methods to be used in its exploration. Later chapters will be devoted to a more detailed working out of these basic concepts in relation to particular classes of polymers.

1

2. What is a polymer?

The word polymer, meaning literally *many parts* (from the Greek *polus*, many, and *meros*, parts, segments), is used to embrace all those materials whose molecules are made up of many units, these units consisting of either a single atom or (more usually) a small group of atoms in a state of chemical combination. An example of a polymer whose unit consists of a single atom is so-called ' plastic sulphur ', with which the reader may be familiar. This is obtained when molten sulphur (at a suitable temperature) is poured into cold water; its structure can be represented by a single chain of atoms joined together by chemical bonds, thus

$$-S-S-S-S-S-S-S- \qquad (1.1)$$

In this state sulphur has entirely different physical properties from ordinary crystalline or *rock* sulphur—properties which are typical of a rubbery polymer. It is soft, highly elastic and translucent. It does not have a definite melting point, like a crystalline material, but on raising the temperature it first softens, then flows like a liquid of very high viscosity. Polymeric sulphur, however, is not stable, and changes back to the familiar powdery or crystalline form after a few days at room temperature.

In most polymers the repeating unit in the structure is a small group of atoms combined in a particular manner. One of the simplest of the polymers, from the standpoint of chemical structure, is polyethylene or *polythene*, in which the repeating unit is the CH_2 group (one carbon atom with two attached hydrogen atoms), these units being joined together to form a long chain, as below

$$-CH_2-CH_2-CH_2-\cdots\cdots\cdots-CH_2-CH_2-CH_2- \qquad (1.2)$$

As its name implies, polyethylene is formed by the joining together of molecules of ethylene

$$CH_2=CH_2 \qquad (1.3)$$

the first stage of the process being equivalent to the ' opening out ', by means of a suitable chemical activator or catalyst, of the double bond, thus,

$$---CH_2-CH_2--- \qquad (1.4)$$

The two outermost single bonds are then able to join up with neighbouring units to form a chain of CH_2 groups in which all the C atoms are connected by single bonds. The original molecule from which the polymer is formed is called the *monomer* unit (Greek *monos*, single). As this example shows, the monomer unit is not necessarily the same as the *repeating* unit in the chain; in the present case the original ethylene unit corresponds to *two* repeating units in the final chain.

Another common polymer of rather similar structure to polyethylene

2

is polypropylene. This is formed by the joining together of molecules of propylene

$$CH_2 = CH \atop | \atop CH_3 \qquad (1.5)$$

to form the chain

$$-CH_2-CH-CH_2-CH-CH_2-CH- \atop | \qquad | \qquad | \atop CH_3 \qquad CH_3 \qquad CH_3 \qquad (1.6)$$

This structure differs from that of polyethylene in having the methyl group (CH_3) in place of one of the H atoms on alternate carbon atoms of the chain. In polypropylene, however, the repeating unit in the chain corresponds to the original monomer molecule.

In natural rubber the repeating unit is rather more complex. It is represented by the formula

$$-CH_2-CH=C-CH_2- \atop | \atop CH_3 \qquad (1.7)$$

and is known as the isoprene unit. This, it will be seen, contains *four* chain carbon atoms, as well as a methyl (CH_3) side group. It is distinguished from polypropylene also by the presence of a double bond in the chain; this has a very important effect on the chemical reactivity of the rubber molecule, and is of fundamental importance in the process of vulcanization (see Chapter 4).

The examples given above will be sufficient to illustrate the distinguishing characteristic of a polymer molecule, namely its chain-like structure, formed by the repeated addition of successive units *end-to-end*. In the examples quoted, each of these repeating units is of identical composition and structure. It is not necessary, however, that all the units in the chain shall be identical. Many polymers, of which one of the principal forms of nylon (called nylon 66) is an example, are formed by a chemical reaction involving two *different* kinds of monomer units or chemical compounds. This results in a structure which may be represented thus:

$$-[A]-[B]-[A]-[B]-[A]-[B]- \qquad (1.8)$$

the individual [A] and [B] units alternating in a regular succession along the whole length of the chain. The final structure may be regarded as having the repeating unit —[A]—[B]—. In other polymers (called co-polymers) the proportions of the two different units [A] and [B] may be varied at will, and the succession of the two units along the chain is usually random, thus:

$$-[A]-[B]-[B]-[A]-[A]-[A]-[B]- \qquad (1.9)$$

3

A number of synthetic rubbers are of this type. Another variant of the same basic pattern is obtained if one of the units, say B, may combine with an A group not only at each end but also at a third point. This gives rise to the possibility of chain branching, as represented below.

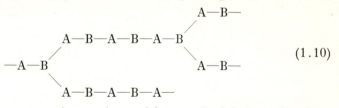

$$(1.10)$$

Such a polymer may continue to ' grow ' from each of the branch points, so building up a complicated highly branched 3-dimensional structure.

Up to this point no attention has been given to the question of the *number* of individual units in the molecule which are necessary if the material is to be classed as a *polymer*. What number constitutes *many*?

There is no precise answer to this question. Strictly, any number from two upwards constitutes a polymer. However, polymers containing only a small number of units are usually called dimers, trimers, tetramers, etc., according to the number of original molecules or monomer units involved, and the term polymer (or *high* polymer, to be more precise) is reserved for the case in which the number of units is very large. In practice the minimum number of monomer units may be taken to be about one hundred. At the other end of the scale, there is theoretically no limit, and polymers containing as many as 100 000 monomer units may be produced. The great majority of polymers, however, fall in the range 1000 to 10 000, corresponding to molecular weights of the order of, say, 14 000 to 500 000, depending on the molecular weight of the monomer unit.

3. *Natural polymers*

(*a*) *Fibres*

Among the natural polymers of industrial significance the most important place is occupied by the fibres, which may be of either plant or animal origin. The major use of fibres is, of course, in clothing and other textile fabrics, but they also have important industrial applications as reinforcing materials in rubbers or other polymers, e.g. in tyres, conveyor belting, etc. The outstanding property of a fibre is its high tensile strength; the fibres are, in fact, among the strongest materials known (see Chapter 9). This particular property arises from the very special arrangement of the molecules in the structure of the fibre. This structure will be considered in detail in Chapter 8. It is sufficient for the present purpose to note that fibres generally contain very small crystals or crystallites, and that these crystallites are lined up or ' oriented ' in such a way that the long-chain molecules all lie in a direction parallel

4

or nearly parallel to the axis of the fibre.   This geometrical disposition of the chains is the most effective arrangement for resisting deformation or breakdown of the structure under the action of a tensile stress.

The suitability of fibres for use as textile materials is associated not only with their mechanical strength, but also with another physical property, i.e. warmth or thermal insulation.   This is a property of the fabric, considered as an assembly of millions of fibres, rather than of the fibres themselves.   The original fibres have to be drawn together and twisted or ' spun ' to form a yarn; the yarn is then woven or knitted to produce the final fabric.   These various processes are designed to give cohesion and strength to the finished product while retaining a considerable proportion of air space between the fibres in the yarn, or between the yarns in the fabric.   It is the air entrapped in the structure which is responsible for the low thermal conductivity of the material.   The looser or more ' open ' the structure (within limits) the lower is the thermal conductivity and the warmer the ' feel ' of the fabric.   Strength is required not only in the finished article, but also in order that the fibres shall be able to withstand the very high mechanical stresses to which they are subjected in the many complicated processes involved in high-speed spinning, weaving and knitting operations.

These two properties, strength and warmth, which man finds so valuable, account also for the uses to which fibres are put in the plant and animal world.   Fibres, in the form of hair, provide the natural clothing of animals, while in plants, the main function of fibres is to give strength to the growing structure.   Of the natural fibres, by far the most widely used are those whose basic constituent is the chemical compound known as cellulose.   Cellulose occurs in the cell walls of most plants, and forms the major component of wood.   The most important cellulosic fibres include flax (or linen), hemp and jute, obtained from the stems of plants, and cotton, which occurs as a hairy mass surrounding the seed.   Chemically, cellulose has a rather complicated chain structure, the repeating unit being represented by the formula $C_6H_{10}O_5$.   This unit is closely related to the molecule of glucose, one of the simpler forms of sugar; for this reason cellulose is classified scientifically as a poly*saccharide* (from the Greek *sakkharon*, sugar); its detailed structure is considered in Chapter 8.

Of the remaining industrially important natural fibres, it is only necessary to mention wool and silk.   Both of these are animal products, silk being the filament extruded from a gland by the silkworm to form its cocoon in much the same way as a spider produces the filaments from which it forms its web.   The silkworm is actually a rather large caterpillar (fig. 1.1); it wraps itself up in its silky cocoon before retiring to change over to the chrysalis stage, whence it eventually emerges as the moth *Bombyx mori*.   Wool and silk are known chemically as *proteins*—a very large class of materials which includes a wide variety of plant and animal products, many of which, e.g. the proteins from seeds, such as

5

Fig. 1.1. The silkworm.

peas, beans, corn, etc., as well as meat (muscle fibre) are valuable as foodstuffs. The proteins differ from the polymers already considered in that they do not have a single precise chemical structure. They are based on a chain repeating unit of the general form

$$-NH-CH-CO-$$
$$\underset{R}{|}$$

(1.11)

in which the group R varies in each successive unit along the whole length of the chain, and may be any one of a number of different chemical units, called *amino-acids*, of which there are some twenty-five or more in all. The different proteins are characterized by different selections, and different proportions, of these substituent groups.

(*b*) *Rubber*

The natural rubber of industry is the product of a species of tree known as *Hevea braziliensis*, which, as its name implies, was originally found in Brazil. On cutting through the bark into the underlying system of latex vessels, the rubber *latex* exudes as a milky-looking liquid in which the rubber is suspended in the form of microscopic globules. The rubber industry expanded from small beginnings in the early nineteenth century, and was originally based on trees found growing in the wild state; at the beginning of the twentieth century, however, plantations developed in Ceylon, Malaya (Malaysia) and the Dutch East Indies (Indonesia) began to oust the original 'wild' rubber from Brazil.

A number of trees and plants produce a rubber identical to that obtained from Hevea; one of these is *Ficus elastica*, closely related to the familiar decorative 'rubber' plant. Apart from this there is another tree which produces a kind of rubber known as *gutta-percha*, whose molecule contains the same unit as Hevea rubber (isoprene) in a slightly different structural form.

The outstanding properties of rubber are too familiar to need recounting. From the time of its original commercial exploitation until the present century its high extensibility and resilience placed it in a class by itself—a kind of scientific curiosity—a position which it continued to occupy until the emergence of the concept of a high polymer and the parallel development of synthetic rubberlike materials.

6

(c) Biological polymers

The structure of the animal body makes use of the physical and chemical properties of a variety of polymeric materials. Reference has already been made to muscle, which is made up of bundles of fibres, and is a form of protein. The main function of muscle is, of course, to convert the chemical energy derived from food into mechanical work, but it also possesses some of the elastic properties of rubber, and the muscular covering of the body provides a resilient cushion to absorb shocks and protect the internal organs from damage. Glue or gelatin is obtained from another fibrous protein (collagen) which is the main component of skin. Collagen is also found in tendons (which connect the muscles to the bones), ligaments, etc., and it is an important constituent of bone itself. The toughness of leather, which is produced by the chemical treatment (tanning) of skin, is due to the network of collagen fibres from which it is built up.

4. Synthetic polymers

(a) Fibres

Among the fibres, we have to distinguish between the true synthetics, i.e. materials whose large molecules are built up or *synthesized* from very simple chemical compounds, and those which are formed by the conversion of a naturally occurring polymer (usually cellulose) into a different form by chemical processing. Both these types are included under the general heading ' *man-made* ' fibres.

The earliest forms of artificial fibres were the result of attempts to imitate the smoothness and beauty of natural silk, and were known as ' artificial silk '. The silkworm produces a single continuous filament, about 1 km in length, which has to be subsequently unwound from the cocoon and ' spun ' to form a yarn or twisted bundle of filaments. The smoothness of the yarn arises from the great length of the filaments of which it is composed. Cotton yarns, on the other hand, are made from short fibres (of length about 25 mm); the filaments in the yarn are imperfectly aligned, and their ends protrude to give a hairy or rough feel. The objective in producing ' artificial silk ' was to obtain a material which could be produced in the form of a single continuous filament, like that provided by the silkworm.

For the production of a fibre in continuous-filament form the starting material must be in the liquid state, i.e. in the form of a molten polymer or polymer solution. In considering cellulose as a possible material for such a development, the immediate difficulty which has to be overcome is that cellulose does not melt, nor is it soluble in water or in any of the usual solvents. In order to make use of it, therefore, it has to be chemically treated or modified in some way. One process for accomplishing this consists in reacting it with acetic acid, whereby the original cellulose is converted to a different material, cellulose acetate. This

7

reaction can be carried out without affecting or breaking down the basic chain structure in any way; the polymeric form of the molecule is thus preserved.

Cellulose acetate dissolves readily in organic solvents, e.g. acetone, to form a highly viscous or syrupy solution. The solution may be pumped through a die, or *spinneret*, containing a suitable number of very small holes, to produce an assembly of fine filaments which, on drawing out and drying off of solvent, form a continuous-filament yarn of cellulose acetate. In another form of process the extruded liquid jet or thread of a chemically modified cellulose (in this case a sulphur-containing compound, cellulose xanthate) is subjected to a further treatment which converts it back to the original cellulose. This product, known as viscose rayon (from the viscous solution from which it is produced), is an example of a *regenerated* cellulose fibre.

The true synthetic fibres are all produced in continuous-filament form. These materials—unlike cellulose—can usually be melted without difficulty, and the process of extrusion may therefore be carried out at a temperature above the melting point of the polymer rather than from solution. The synthetic fibres are not to be regarded as substitutes or *artificial* materials; they have properties different from, and in certain respects superior to, any of the natural fibres. There are many types of synthetic fibre, marketed under a variety of trade names; the structure of some of these will be considered in detail in a later chapter. For the present we shall only note that they include such materials as the nylons, the polyesters (Terylene), the acrylic fibres (Orlon) and the recently developed polypropylene fibre (Ulstron).

In natural fibres the necessary alignment or arrangement of the polymer molecules along the direction of the length of the fibre is laid down in the actual process of growth. In the synthetic fibres, on the other hand, the filaments as originally produced, e.g. from the melt, have little or no molecular ' orientation '; the molecules are not lined up parallel to the axis of the fibre. Such filaments are comparatively weak, and are quite unsuitable for use as fibres. In order to produce the required molecular orientation they have to be subjected to an additional stretching or drawing operation. The conditions under which this is carried out (temperature, rate of extension, etc.) have an important effect on the final structure of the fibre, and hence on its ultimate strength. This question will be considered in more detail in Chapter 8.

(b) Rubbers

Synthetic rubbers were first produced in Germany, a few years before the outbreak of World War II (1939–45). Their production was part of Germany's effort to reduce the dependence of her industry on materials imported from abroad. The most important of these synthetic rubbers was Buna rubber, based on the unit of butadiene (pronounced buta-di′-ene)

$$-CH_2-CH=CH-CH_2- \qquad (1.12)$$

a compound having a structure rather similar to that of the isoprene unit of natural rubber (1.7), except that the CH_3 group on the side of the chain is absent.

On the loss of the major sources of supply of rubber from the Far East following the Japanese entry into the war in 1941, Great Britain and her allies were faced with a crisis of the first magnitude. Fortunately supplies of natural rubber from Ceylon were still available; these alone amounted to as much as Britain's total pre-war consumption of rubber. But they were quite inadequate to sustain the total war effort. Africa was searched for rubber-producing plants of all kinds, but although quite a number were found, their potential contribution was insignificant. The situation was rescued by the development of the American synthetic rubber known as GR–S. This is of more complex structure than the German Buna rubber; it is a *co-polymer* of two ingredients, of which the first is butadiene and the second styrene, whose constitution will be considered later. The American development of this butadiene–styrene rubber (based on raw materials from natural gas and petroleum) was one of the most impressive industrial achievements ever accomplished; by the end of the war the amount produced approached the output of the whole of the pre-war natural rubber industry. This synthetic rubber was inherently not quite as good as natural rubber—its poor ' tack ' or adhesion led to difficulties in tyre building, for example—but this and other problems were largely overcome by blending it with a small proportion of the still remaining natural rubber.

Another important synthetic rubber is *butyl* rubber, a polymer produced from isobutylene. Butyl rubber has the property that air (and other gases) diffuse through it very much more slowly than through natural rubber; it is therefore valuable for the manufacture of inner tubes for tyres. Compared with natural rubber, however, its elastic properties are rather poor, particularly at low temperatures, and its uses are consequently somewhat restricted.

In contrast to fibres, rubbers are not normally crystalline; their molecules are not arranged in any sort of order (cf. fig. 4.1, p. 45). This type of structure, which is in many ways similar to the structure of a liquid, is called *amorphous* (Greek *a*, without, *morphe*, form). It is this rather loose structure (as distinct from the tightly bonded and regular structure of a crystal) which gives rubbers their softness and flexibility. This aspect of the subject will be discussed more fully in Chapter 3.

(c) Crystalline polymers

The third class of synthetic polymers—and in many ways the most interesting—is the class of crystalline polymers. These are not wholly crystalline, like an ordinary crystalline solid, but contain a large number of very small crystals existing in close proximity to the remainder of the material, which is in the disordered or amorphous state, as represented in

9

fig. 6.4a, p. 86. The fibres form a special class or sub-group among the crystalline polymers, but whereas in fibres the crystallites are aligned or oriented in a roughly parallel arrangement, the crystals in ordinary crystalline polymers do not have any preferred orientation; their directions are completely random.

Crystalline polymers are a comparatively recent development. In the unoriented form they do not have any natural counterparts; they have properties which were not previously obtainable in any known material.

One of the most generally used and versatile of the crystalline polymers is polythene, whose structure was referred to earlier in this chapter. Discovered in Britain a few years before the war, polythene achieved rapid development and numerous applications during and after the war. One very important field of application is in the electrical industry; it is a superb insulator (or *dielectric*). This property, combined with lightness and flexibility, makes it the ideal material for use in co-axial cables for high-frequency use, and it was exploited extensively in radar applications for war-time purposes. In the form of thin sheets, the toughness and transparency of polythene are important for packaging, for horticultural purposes (greenhouse linings, etc.) and for many other applications. It can also be pressure-moulded and ' blow-moulded ' to form bottles and household utensils as well as industrial components for a wide variety of purposes.

Polythene has one drawback—it melts at a rather low temperature (110–130°C). The more recently developed polypropylene (formula 1.6) which, as we have already seen, is structurally very similar to polythene, has the advantage of a somewhat higher melting point (170°C) both in the unoriented state and in the oriented or fibre form. Otherwise it is very similar in properties to polythene, and may be used for similar purposes. Another important crystalline polymer is nylon, which, though developed originally and still used mainly on account of its excellent fibre-forming properties, may also be produced in bulk form for the manufacture of moulded articles. The melting point of nylon is 265°C.

An interesting crystalline polymer is obtained by replacing all the hydrogen atoms in polyethylene by fluorine atoms, the resulting structure being of the form

$$-CF_2-CF_2-CF_2-CF_2-CF_2- \qquad (1.13)$$

This polymer, known as poly-tetra-fluor-ethylene (P.T.F.E.), has a still higher melting point, namely 360°C. It also has an extraordinarily low coefficient of friction, which makes it suitable for use as a bearing material in machines. The reason for this low friction is not known.

There are many other crystalline polymers, and their number is likely to increase. The above examples, however, will serve to illustrate their main properties and uses. These properties place the crystalline polymers, as a class, somewhere between the rubbers, which are soft and

10

highly deformable, and the glasses, which are hard, non-resilient and brittle. The crystalline polymers are moderately deformable and tough, yet sufficiently hard to retain their shape under moderate stresses. The fibres are, strictly speaking, a special sub-group of the crystalline polymers, but because of their very great importance, and the fact that, historically, they preceded the synthetic crystalline polymers, it is convenient to consider them as a separate class.

(d) Glasses and resins

Glasses are recognized by their high optical transparency, and by their brittleness. Their transparency is a result of the fact that they are not crystalline. As in a rubber, the molecules in the glass are not arranged in any geometrical order; the structure is disordered or amorphous. A single crystal, such as quartz or diamond, may have the optical transparency of a glass, but, in general, crystalline materials are not found in the form of isolated single crystals, but as agglomerates or assemblages of large numbers of small crystallites. Just as the whiteness of snow is due to the reflection of light from the many surfaces of tiny crystals of ice, so the milky-white appearance of a crystalline polymer such as polythene, or of a polycrystalline solid like paraffin wax, is due to the scattering of light at crystal boundaries. In the amorphous structure of a glass, as in a liquid, there are no discontinuities or differences of geometrical arrangement of the molecules from one point to another, and hence no internal boundaries from which light can be scattered or reflected. Provided, therefore, that the molecule itself does not absorb light, such materials are transparent. Among the better-known of the transparent glassy polymers are polystyrene, perspex and polyvinyl chloride (p.v.c.).

It is a remarkable fact that, despite the extreme differences in mechanical properties between a rubber and a glass, these two classes of materials are structurally closely similar. In a polymeric glass the long-chain molecules show the same amorphous or completely disordered type of arrangement as the molecules of a rubber; indeed from the purely geometrical standpoint there is no way of distinguishing the one structure from the other. The important difference lies not in the geometrical arrangement but in the *forces* exerted by one molecule on another. In a rubber these forces are weak, and do not prevent one molecule (or segment of a molecule) from moving away from a neighbouring molecule (or segment); the molecules can therefore move about fairly freely. In a glass, on the other hand, the forces between neighbouring molecules are much stronger and the molecules are effectively bound together as a rigid mass. This fundamental difference will be discussed in later sections of the book, when we come to consider in more detail the properties of rubbers and of glasses, and the relation between the rubbery and glassy states.

It is not essential that a glass shall be transparent, and among polymers there is a wide range of materials with mechanical properties similar to

11

those of a glass, which, however, do not possess the same degree of optical transparency as perspex or polystyrene. These materials are known generally as synthetic resins. Some of these synthetic resins were originally produced before the end of the nineteenth century, though at the time of their discovery their polymeric nature was not appreciated. One of the first to achieve industrial importance was bakelite, named after its discoverer Baekeland, who patented it in 1907. Bakelite is a material of brownish appearance, which was (and still is) extensively used as an electrical insulator.

The method of production of a synthetic resin naturally depends on the nature of the compounds from which it is built up. Bakelite itself is the product of the reaction between two components, phenol and formaldehyde, which link up to form chain-like structures containing alternate phenol and formaldehyde units. There is, however, an important difference between the structure of any of the synthetic resins of this type and the structure of the transparent glassy polymers. In the glassy polymers the individual molecule is a single chain of very great length; these materials are referred to as *linear* polymers. The structure of a synthetic resin is not like this, but consists of rather short segments of chains with numerous branches, of the type indicated in the formula (1.10). In the process of formation of the polymer individual branches may develop in such a way that two growing chains may join up to form a closed loop, as at A and B (fig. 1.2), or a growing branch may react at some point on an already existing chain, as at C and D. The result of such reactions is to produce a highly branched 3-dimensional network structure of an irregular character in which the segments of chains between branch points are comparatively short and contain only a few monomer units. Strictly speaking, such a structure is not made up of individual polymer molecules; the whole structure may be regarded as one gigantic molecule.

Polymers having the highly branched type of structure depicted in fig. 1.2 are invariably brittle and glass-like in mechanical properties. This, however, does not prevent them from finding many practical applications. They possess the great advantage over the crystalline polymers of being unaffected by heat; being highly cross-linked and non-crystalline they do not soften or melt on heating, as do the crystalline polymers, and they are also highly resistant to chemical attack. The effects of their inherent brittleness can be substantially reduced by incorporating suitable fillers, or by reinforcement with paper, fibres, or other materials. Bakelite itself is frequently compounded with ' wood-flour ' which both cheapens the product and enhances its properties (an unusual combination!). It is also possible to incorporate dyes and pigments, and thus to produce materials of bright colour and attractive appearance. Coloured articles in urea-formaldehyde and melamine-formaldehyde resins, for example, are very popular for the production of toys, tableware, and a great variety of household equipment.

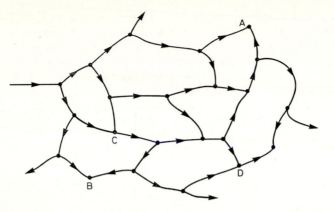

Fig. 1.2.   Structure of synthetic resin of the phenol-formaldehyde
(bakelite) type.

(e) Inorganic glasses

Before leaving the subject of glassy materials, some reference should be made to the relation of the glassy polymers to ordinary glasses, such as window glass.   These glasses are based on mineral or *inorganic* materials, of which the most important are silica (SiO_2) and boric oxide (B_2O_3), together with the oxides of sodium and calcium.   Ordinary window glass, for example, is produced from silica (sand), calcium oxide (lime) and sodium oxide, and is called *lime-soda* glass.   These inorganic glasses are not true polymers in the sense of possessing permanent molecules of chain-like form, but they do form a kind of temporary network structure like that shown in fig. 1.2 in the process of cooling down from the molten state.   They are therefore closely akin to polymers, and they owe their glassy properties to the same basic factors as those which are responsible for the glassy properties of the polymeric glasses.   We shall have to consider what these factors are when we come to examine the nature of the glassy state (Chapter 5).

5. Conclusion

This brief survey of the principal types of polymers reveals the great variety of materials encountered under the general heading of polymers, and the correspondingly great variety of physical properties displayed by them.   At the one extreme we have the rubbers, materials which are soft, deformable and tough; at the other extreme the glassy polymers, which are hard and brittle.   Between these two extremes lie the crystalline polymers, which are moderately hard, yet flexible and strong.   The wide group of materials classed as fibres is a special type of crystalline polymer in which the long-chain molecules are lined up parallel to the direction of the fibre axis.

13

The great scientific advance which the present century has witnessed is the recognition that all these various materials are related by a common basic structure—a structure in which the essential element is the long-chain molecule. The great variety of physical properties which the various polymers display is the result of variations in the chemical constitution of the polymer molecule, which determines the manner in which the individual molecules are arranged or joined together to form the final structure. The precise way in which this comes about forms the basic subject-matter of polymer science, and will be considered under appropriate headings in the following chapters.

CHAPTER 2
the polymer molecule: its size and form

1. *Introduction*

IN Chapter 1 we took a brief look at the various types of polymeric materials and saw that despite their great diversity of physical properties, there was one feature common to them all—a structure of long chain-like molecules. The objective of the polymer scientist is to explain the physical properties of the various types of polymers in terms of the nature of the molecules of which they are composed and the forces by which they are held together. In order to achieve this objective it is necessary first to find out as much as possible about the size and ' shape ' of the polymer molecule itself. This is the basic question to be considered in the present chapter. This will lead on naturally to the more detailed study of particular classes of polymers, starting from rubbers, which are the simplest, and continuing with glasses, crystalline polymers and fibres.

Polymer molecules vary both in size (i.e. chain length) and in chemical constitution. There is no single object which can be called *the* polymer molecule. Nevertheless it is possible to form a general idea or concept of the polymer molecule which represents the significant features of polymer molecules in general—a sort of average or typical polymer molecule —even though individual molecules may depart in various particular directions from this representative type. The formation of this general concept of the polymer molecule was, as a matter of historical fact, extraordinarily difficult to achieve, and its ultimate acceptance, through the accumulation of overwhelming evidence in its favour, marked the birth of the subject of polymer science as a separate branch of chemistry.

2. *The molecular weight* problem*

Of outstanding importance in this development was the refinement of methods of measuring the molecular weight of chemical compounds of very high molecular weight. The classical methods of measuring molecular weight had been developed from experience with materials of comparatively simple composition and low molecular weight: materials whose molecules contained only a small number of atoms, usually not more than ten or twenty. The structure of such molecules, and hence their molecular weight, could usually be deduced from a knowledge of

---

* The molecular weight is the *mass* of the molecule expressed on a scale on which the mass of the ^{12}C carbon atom is 12 units exactly.

15

the chemical processes by which they were originally built up, and of their reactions with other compounds. In addition to these purely chemical methods, however, there are a number of *physical* methods of determining molecular weights; these generally depend on the properties of the material either in the vapour form, or in solution. In most cases these physical methods have been shown to give the same results as the purely chemical methods, but they depend upon the assumption that the material is uniformly dispersed in the form of single molecules, either in the vapour or in the solution. If the molecules tend to stick together or 'associate' in pairs or in clusters incorrect values of molecular weight are obtained. The same applies if they split up or 'dissociate' as does common salt when dissolved in water.

Polymers cannot be evaporated, hence we will consider only methods depending on the measurement of solution properties. Three types of measurements may be used. These are

1. Osmotic pressure.
2. Elevation of boiling point.
3. Depression of freezing point.

The principle of the osmotic pressure measurements is shown in fig. 2.1. The two chambers A and B are separated by a thin *semipermeable* membrane M. A solution of the material to be examined is contained in A, while B contains the pure solvent. The membrane

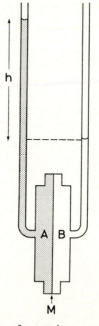

Fig. 2.1. Principle of osmotic pressure measurement.

16

allows solvent molecules to diffuse through it, but is impermeable to the molecules of the solid, or solute. Diffusion of solvent molecules from B to A proceeds until the pressure difference, measured by the difference in level h, is equal to the osmotic pressure of the solution. The type of membrane required depends on the system studied; for many polymers a specially prepared porous cellulose film is the most suitable.

An osmotic pressure measurement really gives us direct information about the *number* of molecules of the solute per unit volume of the solution; it is not *in itself* concerned with the mass or size of the dissolved molecules. But if we know the concentration of the solution, a knowledge of the number of solute molecules per unit volume enables us to calculate the mass of the molecule and hence the molecular weight.

The other two methods, elevation of boiling point and depression of freezing point, both depend on the fact that the vapour pressure of a solution is lower than the vapour pressure of the pure solvent at the same temperature. It can be shown theoretically that the reduction of vapour pressure is directly related to the osmotic pressure. These two methods therefore measure *exactly* the same quantity as the osmotic pressure. The choice of which of the three methods is used is thus dictated not by any theoretical advantage of one over another, but by purely practical considerations. Osmotic measurements are the most sensitive, but accurate results require rather elaborate apparatus and considerable time for equilibrium to be established.

When attempts were made to apply the above methods to the measurement of the molecular weights of materials like rubber, gelatin, starch, etc., the results obtained were wildly inconsistent. In the case of rubber, for example, the osmotic method yielded values of molecular weight in the range 200 000 to 500 000, while the depression of freezing point of solutions gave values of 1000 to 5000. (The boiling point method is not generally used for polymers because polymer molecules tend to break down at high temperatures.) Bearing in mind the theoretical equivalence of these methods, such extraordinary divergences were quite unintelligible. For many years there was a natural reluctance to accept the higher figure, which seemed totally irreconcilable with all existing chemical knowledge. The apparently enormous molecular weight might have been due to some form of physical association or agglomeration of the ' real ' molecules. But then, nobody ever succeeded in isolating or identifying the postulated ' real ' molecule.

Viscosity of solutions

To add to the confusion, a further line of evidence on the value of the molecular weight was provided by the interpretation of the *viscosities* of polymer solutions. This evidence is less direct than the evidence obtained from the osmotic pressure and freezing-point depression methods, and is concerned more with the *dimensions* or size of the

17

molecule rather than with its molecular weight as such. However, if the polymer molecule is in the form of a long chain, there is a direct relation between its dimensions and its molecular weight.

It is a common feature of practically all types of polymers that their solutions have very high viscosities compared with low-molecular materials at the same concentration. Familiar examples are glue (in water) and rubber (in benzene), which yield solutions which are so sticky as to be difficult to pour at concentrations in the neighbourhood of only 5%. Materials having this property were originally known as *colloids*, meaning ' glue-like ' (Greek *kolla*, glue). The basic reason for the high viscosity can be understood by considering the hindrance to the flow of the solvent introduced by the presence of the molecules of the solute. In fig. 2.2 (*b*) the arrows indicate the velocity gradient in a liquid which is being subjected to shear flow* between a fixed plate A and a

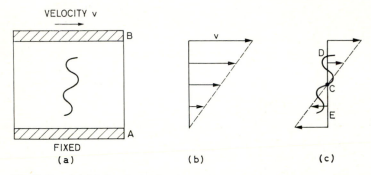

Fig. 2.2. (*a*) Molecule in flowing solution. (*b*) Velocity gradient. (*c*) Velocity of liquid relative to molecule.

plate B moving at constant velocity *v* in the direction indicated. A molecule of the polymer will be carried along by the flowing liquid, and will move with the same average velocity as the liquid in which it is immersed. Owing to the great length of the molecule, however, the velocity of the liquid at one point along its length will be different from that at another point. Thus, if we allow the centre C of the molecule to move with the velocity of the liquid in its immediate vicinity, the liquid at its upper extremity D will be moving faster, and that at its lower extremity E more slowly than the molecule itself (fig. 2.2 (*c*)). These local differences of velocity introduce a frictional resistance which impedes the free flow of the liquid and hence increases the viscosity of the system. Clearly the effect becomes greater, the greater the physical dimensions of the molecule. For a molecule of roughly spherical form the difference of velocity between its upper and lower boundaries would be slight, and the corresponding effect on the viscosity very small. But

* See Chapter 11 for discussion of flow.

18

for a molecule of the same *molecular weight* in the form of an extended chain the effect may be very considerable.

The relationship between viscosity and molecular dimensions was most extensively studied by Staudinger in Germany. From measurements of the viscosities of solutions of molecules of comparatively short chain length whose molecular weight had been obtained by independent chemical methods, Staudinger was led to the conclusion that for a chain-like or *linear* polymer the increase of viscosity of the solution, compared with that of the pure solvent, was directly proportional to the length of the chain. Application of this rule to high polymers such as rubber yielded values of molecular weight of the same order of magnitude as those obtained from osmotic pressure measurements. The viscosity method does not have the same absolute theoretical justification as the osmotic pressure or freezing-point depression methods; nevertheless Staudinger's arguments, which could not be ignored, played an important part in reinforcing the case for the acceptance of the osmotic values of molecular weight rather than the lower figures, and proved ultimately to be substantially correct.

The discrepancy between the values of molecular weight obtained from the osmotic pressure and freezing-point depression methods, which appeared so baffling at the time, is now understood. The explanation is in fact very simple. It has nothing to do with the methods of measurement *as such*, but is due to the differences in the *concentration* of the solution at which the different types of measurement were made. Both methods, as we have seen, essentially evaluate the *number* of molecules in the solution. Now for a given concentration of solution, expressed as mass of solute per unit volume of solution, the number of molecules is inversely proportional to the molecular weight. A polymer of molecular weight 500 000 for example, will therefore give only one thousandth of the freezing-point depression given by a material of molecular weight 500 at the same concentration by weight. The effect of this is that in order to obtain a measurable freezing-point depression in the case of a polymer it is necessary to go to much higher concentrations than would normally be used, namely to concentrations of about 10% by weight. Now the assumption underlying the theory on which the method is based is that the individual molecules act as separate and independent entities. This assumption is only justified if the space available to each molecule in the solution is greater than the dimensions of the molecule itself (fig. 2.3 (*a*)). A simple calculation shows that for a molecule of molecular weight 500 000 this will only be true for concentrations of less than 0·1%, i.e. less than one hundredth of the concentration actually used. At higher concentrations the molecules become entangled with one another, as shown in fig. 2.3 (*b*), and the results are meaningless.

The osmotic method, on the other hand, is much more sensitive, and quite measurable values of osmotic pressure may be obtained at concentrations as low as 0·1 or even 0·01%. This method therefore gives

19

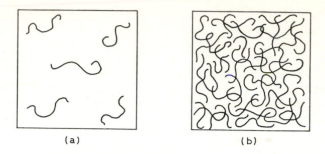

Fig. 2.3. Polymer molecules in (a) dilute and (b) concentrated solution.

completely reliable results in the range of concentrations normally employed.

A further source of difficulty lay in the observation that molecular weight measurements on any given type of polymer varied considerably from one sample to another. To the chemist at the time it was almost a definition of a pure chemical compound that it should have a single and definite molecular weight. The idea that the molecular weight of a substance might vary, or that a single substance might contain molecules of different molecular weights, was not acceptable and such variations were attributed to a failure to isolate a *pure* compound.

Gradually, however, the picture became clearer, and the various difficulties and inconsistencies began to be eliminated by more careful experimental work and a better understanding of the theoretical basis of the observations. By about 1930 the present view, that a polymer is a material containing molecules of very great length, and that this length need not be the same for all the molecules of a given type, began to be accepted. This conclusion, once established, appeared perfectly natural, and on looking back it is difficult to appreciate the strength of the opposition to its acceptance.*

3. *The size of the polymer molecule*

Having considered what is known about the molecular weight of a polymer, we now have to translate this knowledge into terms of geometry. What does it imply regarding the size and form of the polymer molecule?

We begin by considering the question of size. It makes very little difference which polymer molecule we consider; for purposes of illustration we may therefore choose the simplest type of structure, namely the polyethylene molecule

$$-CH_2-CH_2-CH_2-CH_2-CH_2- \qquad (2.1)$$

* Staudinger's wife writes: ' My husband encountered opposition in all his lectures. Only in 1929 when in a lecture. . .he put forward his viscosity formula, for the first time there was no opposition. This both astonished and pleased us.'

20

The 'backbone' structure of this molecule is a chain of carbon atoms joined together by single 'valence' bonds making an angle of $109\frac{1}{2}°$ (the so-called 'tetrahedral' angle)* to each other. To each carbon atom are attached the two hydrogen atoms, also at the tetrahedral angle (fig. 2.4); each carbon atom is therefore at the centre of four tetrahedrally disposed bonds. (This tetrahedral disposition of bonds is most simply seen in the

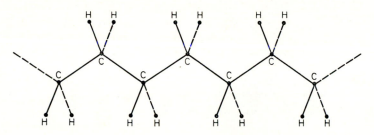

Fig. 2.4. Structure of polyethylene chain.

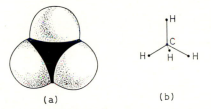

(a)   (b)

Fig. 2.5. Models of methane (CH_4). (a) Space-filling. (b) Skeleton.

molecule of methane (fig. 2.5 (b)). Taking the 'molecular weight' of the CH_2-group as 14, a molecule of molecular weight 350 000—a typical value for a polymer—would contain 25 000 chain carbon atoms or 75 000 atoms altogether.

It requires some effort of the imagination to visualize what this implies. Merely to write down the formula, as in (2.1), for the whole molecule would occupy 55 full pages of this book! If a model were made on the scale of 2·5 cm to 1 ångström (0·1 nm) the distance between the centres of adjacent carbon atoms would be 3·85 cm and the length of the resultant chain would be 780 m.

The structure depicted in fig. 2·4 is a sort of skeleton structure. This is a useful form of representation for showing up the geometry of the chain, i.e. the bond lengths and bond angles, but it gives a rather misleading impression of the space-filling properties or volume occupied by the molecule. A more realistic impression is given by a model in which

* A regular tetrahedron is a pyramid whose four faces are equilateral triangles. The 'tetrahedral angle' is the angle between lines drawn from the centre of the tetrahedron to any two of its four corners.

the atoms are represented by solid spheres (or portions of spheres). The radius to be assigned to atoms of any given type is found from the separation between the atoms of neighbouring molecules in the solid state; this is easily obtained from the X-ray study of crystal structures. Thus, for example, the radius of the carbon atom in the singly-bonded state is 1·35 ångström; that of hydrogen is 0·9 ångström. Figure 2·5 (*a*) shows a model of methane (CH$_4$) utilizing these atomic radii; the impression given is very different from that conveyed by the 'skeleton' structure (*b*). A corresponding model of a section of the polyethylene chain containing 10 carbon atoms is shown in fig. 2.6. This model

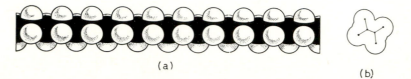

(a)

(b)

Fig. 2.6.   Segment of polyethylene chain containing 10 carbon atoms.
(*a*) Side view.   (*b*) End view.

shows the surface of the molecule to be densely covered with hydrogen atoms, the carbon atoms being so deeply buried that they cannot be separately identified. The cross section of the molecule has a roundish form, with a mean diameter of about 5 ångström. For a chain of molecular weight 350 000, the length is therefore about 6000 times the diameter.

4. *The form of the polymer molecule*

The above illustrations may help the reader to form some sort of mental picture of the geometrical dimensions of a typical polymer molecule. In one respect, however, they are still quite misleading, for they give the impression that the molecule is straight and rigid, like a solid rod of enormous length. The reality is very different—so different, in fact, that it bears very little resemblance to the picture so far presented. A polymer molecule is neither straight nor rigid.

The way this comes about may be seen by considering our model a little more closely. In the skeleton structure depicted in fig. 2.4, the carbon atoms forming the backbone of the chain are represented as all lying in the plane of the paper, with the axis of the chain forming a straight line. But there is no particular reason why they should all lie in this plane; they could equally well be put in various other positions, giving the chain a variety of geometrical forms, or *conformations*, as they are called. These other forms may be brought about simply by allowing the single bonds between successive carbon atoms in the chain backbone to rotate out of the plane of the paper, provided, of course, that we keep the 'valence' angle between each successive pair of bonds unchanged

22

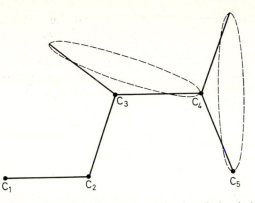

Fig. 2.7. Rotation about successive bonds in chain.

(109·5°). The effect of such rotations is shown in fig. 2.7, in which the first three carbon atoms of the chain may be taken to define the plane $C_1 C_2 C_3$. The fourth atom C_4 must then lie on the rim of a cone formed by the rotation of the bond $C_3 C_4$ about the axis $C_2 C_3$, but it may be *anywhere* on this circle. Similarly the atom C_5 may lie anywhere on the circle formed by the rotation of $C_4 C_5$ about $C_3 C_4$. By repeating this process we arrive at an irregular form of structure (fig. 2.8 (*a*)) which

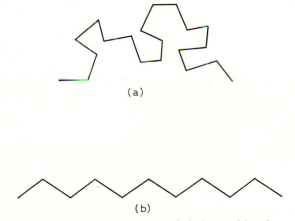

(a)

(b)

Fig. 2.8. (*a*) Irregular conformation of chain resulting from random rotation about bonds. (*b*) Extended form in which bonds lie in single plane.

is geometrically very different from the ' straight ' or regular zig-zag form (*b*). Clearly by choosing different positions for each of the rotatable bonds, it is possible to construct any number of model molecules, each having a different conformation, and there is no reason, in principle, for preferring any one to any other.

23

The justification for the assumption that the individual C–C bonds can rotate with respect to their neighbours is provided by a number of lines of chemical evidence on the geometrical structure of certain types of compounds of low molecular weight. For example, in dichlorethylene (obtained by the substitution of Cl atoms for two of the H atoms in ethylene) it is possible to distinguish between the two forms (*a*) and (*b*) shown in fig. 2.9. It is seen that (*b*) could be produced from (*a*)

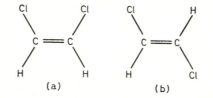

Fig. 2.9.   Alternative structural forms of dichlorethylene.

by rotation of the right-hand CH.Cl group through 180° about the axis of the C=C double bond.   The fact that these two types of structure can be identified shows that *in this case* no such rotation is possible, for, if it were, the molecule would be continually changing from one of these structures to the other, and it would not be possible to isolate molecules of either type.   In dichlorethane, on the other hand, it is not found possible to identify separate compounds corresponding to the types (*a*), (*b*) and (*c*) in fig. 2.10.   (This is a 3-dimensional molecule in which the

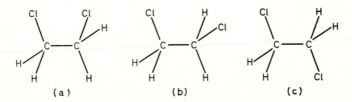

Fig. 2.10.   Possible alternative forms of dichlorethane.

terminal $CH_2.Cl$ group forms a tetrahedron.)   We must therefore conclude that these structures do not have any permanent existence, and since there is no reason for choosing any one in particular, it follows that any given molecule is continually transforming from one to another of these possible forms.   We can thus distinguish between the properties of the C=C double bond, which does not permit of rotation, and the C–C single bond, in which rotation can and does take place quite freely.

A further, even more convincing, piece of evidence is that in molecules containing a chain of six or more carbon atoms and terminated by reactive groups A and B (fig. 2.11) the two ends of a single molecule may join up, by the reaction of A with B, and so produce a ring compound.

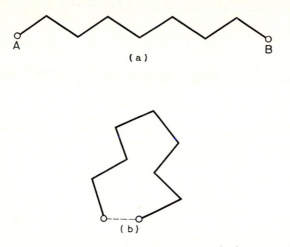

Fig. 2.11.   Ring formation by reaction between terminal groups of molecule.

Such ring formation would not be geometrically possible if the molecule could only exist in the form of a straight rigid chain.

When the molecule is very long, the effect of random rotation about bonds is to produce a highly complicated irregular form, rather like a tangled skein of wool.   This form is illustrated in fig. 2.12, which shows a wire model of a polyethylene molecule containing 1000 C–C bonds. This model was constructed in such a way that the length of each bond was 5 mm and the angle between successive bonds 109·5°, but the rotation of each bond with respect to the plane of the two preceding bonds was selected at random by the throw of a die.   This is illustrated in fig. 2.13.   If A, B and C represent the first three carbon atoms, the fourth atom D may lie anywhere on the circle shown.   Its actual position was determined by the throw of a die, the numbers 1 to 6 on the die giving six equally spaced positions in the circle of rotation, as shown. Thus if the number 3 were thrown, the bond would be placed in the position CD (by suitably bending the wire).   By repeating this process 998 times the structure reproduced in fig. 2.12 was obtained.   The choice of 6 as the number of available positions is of course arbitrary (ideally the number should be much larger), but in practice this gives a very close approximation to completely random rotation.

Bearing in mind that an actual polymer molecule may be very much longer than this—containing perhaps from ten to fifty times the number of atoms shown in the model, we can form some idea of its geometrical complexity.

One further observation may be made here.   In any of the highly kinked-up forms, such as that represented in fig. 2.12, the distance between the ends of the chain is very much less than the actual length of the chain itself—the chain *contour length*—and varies from one particular

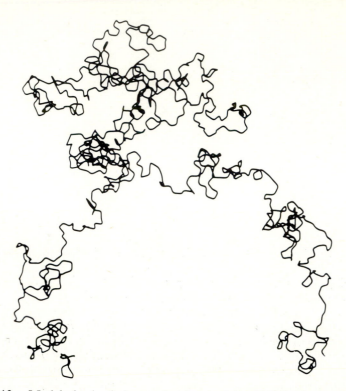

Fig. 2.12.   Model of polyethylene molecule containing 1000 freely rotating C–C bonds.   The positions of successive bonds are chosen at random.

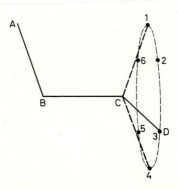

Fig. 2.13.   Selection of positions of successive bonds in circle of rotation.

conformation to another.   In the particular model shown the end-to-end distance was 20 cm.   The chain contour length, however, was 4 m. When we come to consider the elasticity of rubber we shall see that this

26

great difference between the end-to-end distance and the chain contour length is of quite fundamental importance.

5. *Conclusion*

As a result of the various lines of thought referred to in this chapter we are now in a position to form a concept of the size and form of the polymer molecule—a concept which is to a large extent unaffected by the particular chemical features of any given type of polymer. This concept did not arise suddenly; its emergence was the result of the gradual accumulation of knowledge on the structure of molecules of all kinds, and of polymer molecules in particular, and of methods of measuring or otherwise estimating their molecular weight. With its eventual acceptance all sorts of problems, which had previously seemed so puzzling, gradually began to be clarified, and as new observations came along, it was found easier to fit them into the general scientific picture. This is the way in which science advances: the emergence of a new concept, or the introduction of a new theory, is usually preceded by a period of great confusion, in which the antagonists face each other with apparently irreconcilable observations. The new concept, or new theory (if it is successful) cuts right through these conflicts, and a period of rapid development or consolidation usually follows—until the next crisis occurs!

In polymer science we are still being borne along on the wave of success which was set in motion in the early 1930s. The concept of the polymer molecule has proved extraordinarily fruitful and has led on to the understanding of many of the baffling problems which beset earlier workers: the problem of the elasticity of rubber, of how polymers crystallize, of the nature of glassy polymers, and so on. These and other subjects form the main body of this book. There is much that we still do not know; many problems still exist, and new materials—presenting new problems—are continually being discovered. But the outline of an interpretation is now clear, and it rests on the fundamental basis of the nature of the polymer molecule itself—its size and form, and the way in which it interacts with its neighbours.

CHAPTER 3
why is rubber elastic?

1. *Nature of the problem*

IT is well known that the ordinary phenomena of everyday life do not normally arouse our curiosity; they form part of the general background of experience within which our lives run their habitual course. It is the unusual, the unexpected, the exception to the general rule, which attract our attention and stimulate our mental processes.

So it was that rubber, which, when it first began to appear as an article of commerce in the early part of the last century, was regarded as a rather curious material and attracted the attention of a number of eminent scientists, gradually came to be accepted as part of the everyday environment, and hence in no way unusual. Indeed, one may search the textbooks of classical physics and elasticity written in the period from, say, 1870 to 1940 in vain to find any reference whatsoever to this material. Being inexplicable, and not fitting into the classical concepts of ' solids ', it was conveniently dismissed from the mind.

Yet rubber is a fascinating material, possessing some very remarkable properties, the most important of which, of course, is its high elastic extensibility. The question of explaining these properties is a key problem in polymer science, and its ultimate solution marked a turning-point in our whole outlook on polymeric materials—on their molecular structure and physical properties. It is to this problem that we now turn our attention.

First, let us look a little more closely at the nature of the difficulty. Ordinary solids, such as metals, non-metallic crystals, glasses, etc., can be deformed or strained elastically to only a very small extent, corresponding to an extension of not more than about 1%. If an attempt is made to apply a greater extension they either break (as in the case of glass) or deform *plastically*, like steel or (more obviously) lead. A deformation is called *elastic* if the material returns to its original dimensions on removal of the deforming force; if when the deforming force is removed the body remains in the deformed state, it is called *plastic*. Glass is a good example of an almost perfectly elastic solid; it cannot undergo appreciable plastic deformation. Most metals, on the other hand, are capable of a very high degree of plastic deformation; a metal rod, when bent, does not break but deforms plastically.

Rubber is unusual in having the capacity to undergo very large deformations and yet recover its original form on removal of the deforming

28

force. A typical rubber may be extended elastically by as much as 700%, that is to say, to eight times its original length. This is of the order of 1000 times the elastic extensibility of an ordinary solid.

Even more remarkable is the difference in the force required to produce these deformations. A steel wire of diameter 1 mm requires a tension of 1600 N (about twice the weight of a man) to increase its length by 1%; a rubber filament of the same diameter would require a tension of less than 10^{-2} N. In terms of Young's modulus (the ratio of stress to strain), the modulus of steel is about 100 000 times that of rubber.

A moment's consideration is sufficient to show that the very high elastic extensibility of rubber cannot possibly be explained in terms of the classical ideas of the constitution of a solid body. The classical solid owes its 'solidity', or cohesion, to the forces of attraction between neighbouring atoms, which form a closely packed structure. This is illustrated in fig. 3.1, which represents the structure of a simple crystal-

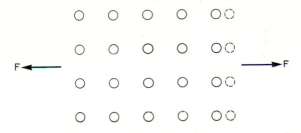

Fig. 3.1. Displacement of atoms in crystalline solid on application of tensile stress.

line solid. Application of tensile forces F to the opposite faces of such a structure causes the planes of atoms to move slightly further apart, as indicated for the outermost layer, shown dotted, and this displacement brings into play corresponding attractive forces between the atoms which exactly balance the applied stress. On removal of the stress these internal attractive forces immediately restore the material to its original unstrained state.

The forces between neighbouring atoms fall off very rapidly as the distance between them is increased (see Chapter 9), with the result that as the deformation is progressively increased a point is reached where they are no longer able to support the stress, and the structure breaks down. This breakdown may be either catastrophic, as in brittle fracture, or more gradual, as in plastic deformation. In either case it is completely irreversible; the original state is not restored when the deforming force is removed. The maximum elastic deformation which is theoretically possible before breakdown occurs is of the order of 10 to 20%. This classical model is therefore quite incapable of accounting for the properties of rubber.

2. The Gough–Joule effect

Quite apart from its elastic extensibility as such, rubber displays a number of other interesting and unusual phenomena. Foremost among these is the Gough–Joule effect. As long ago as 1805, Gough made the discovery that if a strip of rubber, stretched by means of a suspended weight, is heated, it *contracts* in length. On cooling, its length increases. These changes of length are *reversible*; the cycle of heating and cooling may be repeated any number of times. Alternatively, if the sample is held at a constant stretched length, the tension is found to *increase* on heating.

In contrast to this behaviour of rubber, ordinary solids, of course, expand on heating, whether in the stressed or in the unstressed state.

The reversible contraction of stretched rubber on heating was confirmed fifty years later by Joule (famous for his experiments on the mechanical equivalent of heat), and is known as the *Gough–Joule* effect. Since that time it has been amply verified by many observers, and fig. 3.2 shows, for example, the result of an experiment by Meyer and

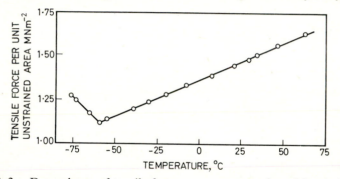

Fig. 3.2. Dependence of tensile force on temperature for rubber held at a constant stretched length (adapted from Meyer and Ferri, 1935).

Ferri, carried out in 1935, in which a sample of rubber was maintained at a constant stretched length, corresponding to an extension of 350%, while the temperature was varied. The tensile force is seen to increase continuously with increasing temperature from −60°C, up to +60°C. (There is a discontinuity at −60°C; below this temperature the effect is reversed. The significance of this is discussed in Chapter 5.)

Demonstration of Gough–Joule effect

The Gough–Joule effect may be readily demonstrated by means of the apparatus shown in fig. 3.3. A rubber band of 15–20 cm circumference is stretched to about three times its original length between the bottom hook A, which is fixed, and the top hook A′, which is connected to a helical steel spring B. This spring is chosen so that it is extended to about twice its original length when the rubber is stretched. A small

30

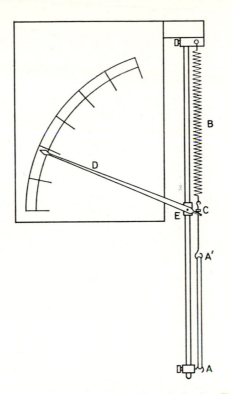

Fig. 3.3.   Demonstration of Gough–Joule effect.

metal disc C soldered on to the top hook rests on the shorter arm of the
light pointer D, which is pivoted at E.   On dipping the rubber into a
beaker of boiling water the contraction extends the spring B, causing the
pointer to rise.   Dipping into cold water reverses the effect.

A more elegant demonstration of the same effect is provided by the
rubber-spoked wheel.   In this (fig. 3.4) a light aluminium rim cut from
3 mm sheet has equally spaced saw-cuts around its circumference, into
which are fitted rubber bands (about thirty is a convenient number)
stretched to about three times their original length.   The wheel, whose
diameter may conveniently be about 25 cm, is mounted on a hub suspended by pivot bearings or ball bearings to eliminate friction.   The
wheel is first carefully balanced by adjustment of the small weights A, B
and C working on screws attached to the rim.   On applying heat from a
small (100 watt) electric heater to one side of the wheel contraction of the
rubber bands occurs; this displaces the centre of mass and causes that side
to move upwards.   As the bands come successively under the influence
of the heater, they in turn contract, thus causing continuous rotation of
the wheel.

31

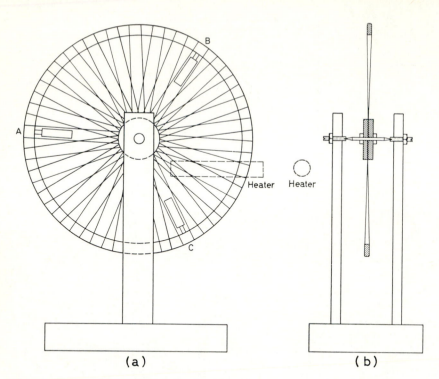

Fig. 3.4. Rubber-spoked wheel. (*a*) Front view. (*b*) Lateral section. Heating causes contraction of the rubber on one side, which throws the wheel out of balance, and results in continuous rotation.

3. *Heat of extension*

Closely related to the contraction of stretched rubber on heating is the development of heat on extension. If a specimen of rubber is quickly extended, its temperature rises. If, after being held in the stretched state until it has reached room temperature, the two ends are moved towards each other again, its temperature falls. The development of heat is therefore *reversible*.

This effect also was discovered by Gough and confirmed by Joule (1859). Joule's result is reproduced in fig. 3.5, together with some recent observations made by James and Guth (1943). It is seen that after an initial very slight cooling the temperature rises progressively as the extension is increased. The total temperature rise at high extensions may amount to as much as 10K (fig. 3.6).

This reversible heat of extension may be easily demonstrated by quickly extending a rubber band to its fullest extension and applying it to the lips, the change of temperature being readily sensed. If, after cooling, it is allowed to contract by bringing the two ends together, it will be

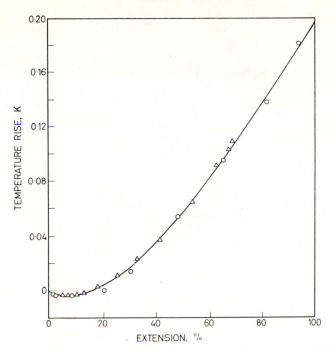

Fig. 3.5. Rise of temperature on stretching of rubber up to 100% extension. ○ Joule (1859). △ James and Guth (1943).

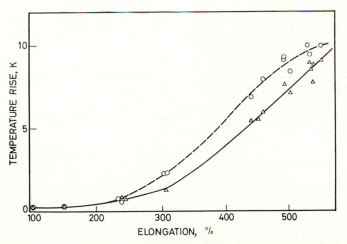

Fig. 3.6. Rise of temperature on stretching of rubber up to highest extension. (Dart, Anthony and Guth, 1942.)

found to be noticeably cooler. (It is important that the rubber be allowed to do work against the applied force during its retraction; if one end is released, allowing the rubber to snap back, kinetic energy is dissipated and no cooling will occur.)

For a more precise demonstration one junction of a fine thermocouple (e.g., copper–constantan) may be inserted into a small hole in the rubber, and the e.m.f. developed indicated on a galvanometer (fig. 3.7).

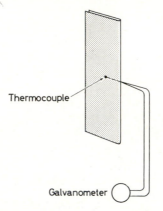

Fig. 3.7. Measurement of temperature rise on stretching by means of a thermocouple.

4. *Early theories of rubber elasticity*

A satisfactory theory of rubber elasticity must provide an explanation not only of its purely mechanical properties, but also of the remarkable thermal or thermo-elastic effects, as they are called, referred to in the previous section. Not unnaturally, the earlier theories, of which there were a large number, concentrated mainly on the purely mechanical properties, and in particular on the high elastic extensibility, which they endeavoured to explain in terms of the properties of the molecules or of other postulated components of the structure.

Although none of the theories put forward before 1932 has survived in its original form it is interesting and instructive to take a quick look at the general lines along which they were evolved, for despite their lack of success, each of them in its way contained a germ of truth and played some part in the developments which led to the ultimate solution of the problem.

Two principles may be utilized to produce a highly deformable structure from a material which, in itself, is capable of only small elastic deformations. One is the principle of the open network, and the other the principle of the coil spring. Most of the early theories of rubber made use of one or other (and sometimes both) of these principles. Very popular at one time were the 2-phase theories, which envisaged an open

34

network structure composed of a rather rigid elastic component immersed in, or interpenetrated by, a soft, liquid-like medium which filled up the spaces of the network but made no contribution to the elastic restoring force. The idea that rubber contained two different components was apparently supported by various observed facts; one was the fact that natural rubber is not completely soluble in solvents such as petroleum, one part—the so-called *sol* rubber—going readily into solution, while another part—the *gel* rubber—either remains undissolved, or dissolves very much more slowly. These two components were believed to be chemically different materials, though their exact constitution was in doubt. It was, however, not unreasonable to suppose, in accordance with these observations, that the insoluble (and more rigid) component was the elastic component of the structure, capable of supporting the applied stress, while the soluble component was more fluid and played the part of a neutral medium separating the elements of the more rigid structure without impeding their movement.

The other class of theories—the molecular-spring theories—became prominent only when more had already been learnt about the structure of the rubber molecule. Starting from the known chemical structure, that is, a given sequence of atoms joined together by chemical bonds of specified type in the form of a long chain, they postulated a ' corkscrew ' type of configuration resembling a coil spring or helix. This configuration was presumed to be maintained by the forces between neighbouring atoms, or between adjacent groups on successive turns of the helix; extension of the molecule along the axis of the helix would result in torsion or twisting of the chain about its own axis which would bring into play attractive forces tending to restore the original form. The theory of Mack published in 1934, which was worked out in considerable detail, was the most promising theory of this kind.

An apparently more far-fetched theory, of a totally different kind, which however was actually very much closer to the truth, was the *skipping-rope* theory proposed by Griffith in 1930. This postulated that the thermal motion, which is possessed by the atoms and molecules of all materials, takes the form, in the case of rubbers, of kinetic energy of rotation of the long-chain molecules as a whole, the motion being precisely similar to that of a skipping-rope with fixed ends. This type of motion produces a force tending to pull the two ends of the rope towards each other (fig. 3.8); the molecule therefore behaves as if it were elastic. For any given amount of rotational energy it is possible to work out a precise relation between the tensile force and the distance between the ends of the rotating molecule.

Fig. 3.8.   Griffith's ' skipping-rope ' theory.

35

Griffith's theory has two considerable advantages over the other theories which have been referred to. Firstly, it enables very large extensibilities to be accommodated without difficulty. Secondly, since the thermal energy of rotation is proportional to the absolute temperature, the tension on the chain (at any given length) should increase in a corresponding manner: the theory therefore provides an immediate explanation of the Gough–Joule effect. Its one drawback—unfortunately a fatal one—is that it fails to indicate any way in which the large amount of free space which would be required for the individual molecules to execute their revolutions is to be provided.

5. *Elastic molecules*

It was Meyer in Switzerland who, in 1932, first proposed the explanation of rubber elasticity which is now generally accepted. Meyer was struck by the similarity in structure between the various materials which, though not so highly deformable as rubber, nevertheless display some degree of rubber-like elasticity, corresponding to extensions of between, say, 50 and 200%. Among such materials may be mentioned wool (particularly in the presence of water), silk, gelatin and muscle fibres. The earlier theories of rubber elasticity had concentrated almost exclusively on rubber itself; they sought to explain this phenomenon on the basis of some peculiar feature of the rubber molecule. Meyer, on the other hand, was impressed by the *generality* of the phenomenon, and took the bold step of attributing it to some factor which was common, if not to all long-chain structures of this kind, at least to a large class of them.

The way this came about has already been hinted at in the last chapter. It was Meyer, in fact, who first clearly recognized that a polymer molecule was not rigid, like a straight rod, but capable of changing its form by virtue of the various independent vibrations and rotations of the individual atoms of the chain, associated with their thermal motion or heat energy. He conceived the idea that this capacity for changing its conformation could—and indeed *must*—lead to the conclusion that the molecule itself possesses a kind of elasticity. For all of the conformations which may arise as a result of these random rotations the great majority will be of the highly crumpled or kinked-up form such as that depicted in fig. 2.12; conformations in which the chain ends are widely separated can arise only by a rather unlikely combination of rotations and will be comparatively rare. In the limiting case, the particular conformation in which the chain is fully extended, like a zig-zag, with its axis lying along a straight line (fig. 2.8 *b*), can arise in one way only, and is therefore very unlikely indeed to occur. Therefore, on probability grounds alone, we can see that if the chain could somehow be pulled out by applying forces to its extremities, it would, on removal of these forces, return in the course of time to one or other of these more highly kinked-up forms. But this, after all, is what we mean by elasticity; it is the tendency of a

36

system to return to some original or normal state after being forcibly deformed. The two ends of the molecule act *as if* there were a force tending to draw them together.

6. *Chain statistics*

Meyer's concept of the origin of the elasticity of the long-chain molecule has been taken up and developed mathematically by a number of authors, notably by Guth and Mark and by Kuhn. From the mathematical standpoint the precise details of the chain structure, or chain geometry, make very little difference to the result. Hence, for the sake of simplicity, it has been customary to carry out the mathematical operations on a kind of abstract or idealized chain consisting simply of a succession of identical links of equal length, the direction of any one link being assumed to be completely random and independent of that of neighbouring links in the chain (fig. 3.9). This model does not accurately repre-

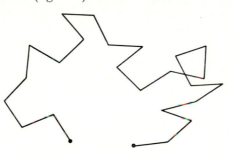

Fig. 3.9. The ' randomly jointed ' chain.

sent the geometry of a real molecular chain, in which successive bonds meet at a certain fixed angle (the valence angle) as shown in fig. 2.7, but this difference is not important. This idealized model of the molecule may be called a *randomly-jointed* chain, or simply a *random* chain.

It is possible by mathematical analysis to treat a chain of this kind *statistically*, that is to say, to derive certain *probabilities* and certain *average* properties. For example, it is possible to find the probability that the two ends of the chain shall be separated by any given distance r; it is also possible to find the *most probable* distance between the ends, and so on. The problem is in fact very similar to the classical problem of the *random walk* (or drunkard's walk, as it is sometimes called) originally treated by Einstein. In this we imagine a man setting out from a given origin and making a succession of steps at random, the direction of any one step being completely independent of that of the previous step. The problem is, where will he get to after n such steps? The answer is of course that we do not know. But we can say this. The probability that the n steps will be all in the same direction, i.e. that the path will be a straight line, is exceedingly small. We can also say that the probability

37

that he will arrive back at the starting point is also very small. These results are obvious. What is not so obvious is that his final distance from the origin is, *on the average*, proportional to the *square root* of the number of steps taken, i.e. to $\sqrt{n}$.

The random chain problem is equivalent to a random walk (or more strictly a random flight) in three dimensions. The mathematical analysis is precisely similar to that for the random walk in two dimensions, and the conclusions are of the same kind. The probability that the distance between the ends of the chain shall have the value r is described by a curve having the form shown in fig. 3.10. This shows that

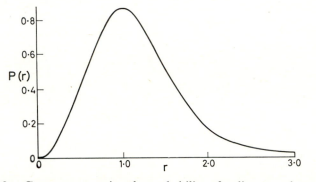

Fig. 3.10. Curve representing the probability of a distance r between the two ends of a randomly-jointed chain (equation (3.1).

both very large and very small values of r are highly improbable; the maximum probability occurs at some intermediate value. This curve is defined by the mathematical formula

$$P(r) = Ar^2 \exp\left(-b^2 r^2\right) \tag{3.1}$$

in which $P(r)$ is the probability of the end-to-end distance r, A and b are constants. The constant A is of no significance and is introduced only to make the total probability equal to unity; the constant b, however, is a function of the chain geometry, and is given by

$$b^2 = \frac{3}{2nl^2}, \tag{3.2}$$

where l is the length of the individual link and n is the number of links.†

It is convenient to define the *root-mean-square* (r.m.s.) length for a chain of this type. (This is obtained by finding the mean value of r^2 and taking the square root; it is mathematically a more convenient quantity

† The reader who is familiar with the kinetic theory of gases will recognize the analogy between the probability function (3.1) and Maxwell's distribution unction, which gives the distribution of velocities among the molecules of a gas, or alternatively, the probability that any given molecule has a particular velocity. It is only necessary to replace the length r in (3.1) by the velocity v.

than the simple average.)   The value for this is found to be a quite simple expression, i.e.

$$\text{r.m.s. length} = l\sqrt{n}. \qquad (3.3)$$

As for the random walk in two dimensions, the r.m.s. length is proportional to the square root of the number of links in the chain.   The *most probable* length r_{mp}, which is defined as the position of the maximum of the curve in fig. 3.10, is closely related to the r.m.s. length, and is given by

$$r_{mp} = (\sqrt{2}/\sqrt{3}) \times \text{r.m.s. length}. \qquad (3.4)$$

The implications of this analysis are very significant.   Whether we take the most probable value or the root-mean-square value of r to represent the normal free or unrestrained state, the result obtained is that this length is proportional to the *square root* of the number of links in the chain.   The outstretched length, on the other hand, is proportional simply to the number of links n.   The potential extensibility of the chain is therefore proportional to $n/\sqrt{n}$, i.e. to $\sqrt{n}$.   If, for example, the chain contains one hundred links, its r.m.s. length will be $l\sqrt{100}$ or $10\,l$ and it will extend to ten times its normal (r.m.s.) length.   If it contains 100 000 links its extensibility will be one hundred times, and so on.   This shows at once that it is only materials in which the chain length is very high that high extensibilities are to be expected.

It is not difficult to adapt the above statistical treatment to take account of the actual geometrical structure of any real molecular chain.   The result is simply to modify the value of the constant b in equation (3.1), without affecting the *form* of this equation.   This modification therefore affects the numerical value of the r.m.s. length but in no way alters the general mathematical relations.   For example, in the case of a chain such as the paraffin or polyethylene chain, in which successive bonds are connected together at a definite angle (called the 'valence' angle, since it is determined by the chemical or valence bonds of the structure), the r.m.s. length is given by the formula

$$\text{r.m.s. length} = l\sqrt{n}\,\sqrt{\left(\frac{1+\cos\theta}{1-\cos\theta}\right)}, \qquad (3.5)$$

where θ is the supplement of the valence angle (fig. 3.11).   Taking the usual value of θ (70·5°) we have $\cos\theta = \tfrac{1}{3}$ and this reduces to

$$\text{r.m.s. length} = l\sqrt{2n}. \qquad (3.6)$$

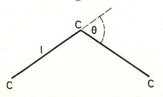

Fig. 3.11.   Valence angle structure.

39

The effect of the valence angle is thus to increase the r.m.s. length of the chain in the ratio $\sqrt{2} : 1$ compared with a random chain having the same number of links.

7. Thermo-elastic effects

In the foregoing sections we have treated the kinetic or statistical theory of rubber elasticity as a necessary, and indeed almost self-evident, consequence of the known facts regarding the nature of the long-chain molecule. However, a theory does not become scientifically accepted on the basis of inherent plausibility alone; it must also lead to a satisfactory description—and preferably a quantitative description—of the main experimental observations. In the case of rubber these observations are concerned with both the mechanical and the thermal, or thermo-elastic properties.

With regard to the mechanical properties, the statistical theory may be further developed to give precise quantitative relations between the stresses and strains when rubber is subjected to a specified deformation, and these stress–strain relations may be compared with experimental results. This aspect of the subject is dealt with in the following chapter. In the present chapter we shall consider the thermo-elastic properties, namely the Gough–Joule effect and the reversible heat of extension.

The significance of these effects was fully recognized by Meyer, who showed that they could be understood as a simple and direct consequence of the kinetic theory.

As has already been noted in connection with the Griffith ' skipping-rope ' theory, any theory which attributes the elasticity to the *thermal energy* of molecules (or of atoms within the molecule) automatically leads to the conclusion that the elastic restoring force (at constant extension) should be proportional to the absolute temperature. The origin of the elasticity is *kinetic*, not static as in the classical theory of elasticity. Since the kinetic energy of the atoms or molecules is directly proportional to the absolute temperature, all the effects associated with this kinetic energy (in this case the elastic force) vary in a similar manner.

There is indeed a very close parallel between the kinetic theory of rubber elasticity and the well-known kinetic theory of gases, and Meyer himself drew a comparison between the elasticity of rubber and the elasticity of a gas. A gas, of course, does not have a shape, but it possesses elasticity with respect to volume. In order to reduce the volume of a given mass the applied pressure must be increased, and on reducing the pressure again the original volume is restored. The pressure, at constant volume, is proportional to the absolute temperature that is, the temperature on the Kelvin scale (Charles's law). Furthermore, the compression of a gas (in which work is done on the gas by the applied force) is accompanied by the evolution of heat, just as the extension of rubber is accompanied by the evolution of heat.

40

This similarity in properties is a reflection of the close similarity of the physical processes involved in the two cases. The pressure exerted by a gas arises from the bombardment of the walls of the containing vessel by its individual molecules, and is proportional to their average kinetic energy, i.e. to the absolute temperature. Elementary kinetic theory shows that for a monatomic gas, at the absolute temperature T (K), the average kinetic energy per molecule is $\frac{3}{2}RT/N$, where R is the universal gas constant for 1 mole, and N the number of molecules in 1 mole. Boltzmann's constant k is R/N, so this is usually written $\frac{3}{2}kT$. Similarly the tension in stretched rubber is associated with the local bombardment of the atoms of the long-chain molecules; the kinetic energy involved in this process has the same form of temperature dependence. For an *ideal* gas there are no forces exerted between the molecules except during molecular collisions, so the only energy which such a gas can contain is stored in the form of kinetic energy of its molecules. When work is done on the gas the additional energy supplied therefore appears in the form of additional molecular kinetic energy, i.e. heat. This is why the heat energy of a gas increases when it is compressed, i.e., the temperature rises. For an exactly similar reason rubber evolves heat when it is stretched: the work performed in stretching it is transformed into more intense molecular motion.

8. *Tension on single chain*

The above argument, if carried a stage further, may be used to derive an expression for the tension on a single long-chain molecule. Let us imagine such a molecule to be held with its ends fixed at two points A and B (fig. 3.12). We have seen that the local bombardments of individual

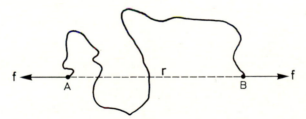

Fig. 3.12. Tension on single chain whose ends are held at a fixed distance apart.

atoms of the chain by the atoms of surrounding molecules will act in such a way that they will tend to reduce the distance between its ends. If this tendency is prevented, the effect will be to create an inwardly directed force on the points of constraint. To counterbalance this force an equal and opposite force, i.e. a tensile force, must be applied to the ends of the chain.

By the application of the statistical theory it is possible to calculate the way in which this tensile force depends upon the distance r between the ends of the chain. The calculation makes use of what is called statistical thermodynamics, and is similar in principle to the calculation of the pressure exerted by a gas on the walls of the containing vessel. The result is expressed in the form

$$f = 2kTb^2r, \qquad (3.7)$$

where k is Boltzmann's constant, T is the absolute temperature in Kelvin and b is the parameter which occurs in the probability function for the chain (equation (3.1)). It is seen that the tension on the chain is proportional to the distance between its ends (fig. 3.13); the molecule therefore behaves like a little spring which obeys Hooke's law (stress proportional to strain), but the tension becomes zero only when the two ends of the chain are brought into coincidence ($r = 0$). For a given value of r the tension is, as expected, proportional to the absolute temperature.

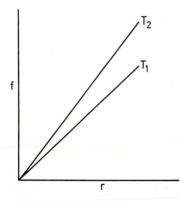

Fig. 3.13. Dependence of tension on distance r between ends of chain.

The tension represented by the formula (3.7) is really only the *average* value of the force; the actual tension fluctuates from moment to moment in response to the random impacts on the chain. The longer the chain, the smaller these fluctuations become. This situation is no different from that to which we have become accustomed in the kinetic theory of gases; the pressure on the walls is continually fluctuating, but the larger the number of atoms considered, the smaller (relatively) are these fluctuations.

9. Conclusion

The great merit of the statistical theory is that it explains the elasticity of rubber in a perfectly natural way, without the necessity for introducing any peculiar special properties related to the chemical structure of the molecule. The fundamental assumptions which it is necessary to introduce are concerned only with the general nature of the chain, the existence of rotation about bonds, etc., and have been established through the accumulation of a great body of chemical knowledge gained from the study of a wide variety of chemical compounds, not specifically related to rubber. It is this *generality*, and the essential simplicity of its underlying concept, which makes the theory so attractive.

There is one sense, however, in which this very generality must be regarded as a source of difficulty. In all the foregoing explanation it has been tacitly implied that *any* molecular structure of the type considered will have properties of much the same kind as rubber; that rubber-like elasticity would be expected to be a general property of all long-chain molecules.† This, as we know, is far from being the case. It is indeed true that all such molecules are *potentially* rubber-like; but whether this potentiality is realized depends on additional factors which have yet to be considered. Before the theory can be regarded as realistic it is as important to understand why in some cases rubber-like elasticity does *not* appear as it is to know why in other cases it does. The modifications of the original concept which have to be introduced to bring about this understanding will be considered in the following chapter.

† An early article by the author in which the elasticity of the rubber molecule was attributed to random rotation about bonds drew the question from a reader, ' If this is the case, why is not the paraffin molecule also elastic? ' (see *J. Soc. Chem. Ind.*, 1943, **62**, pp. 326, 351, 435.)

CHAPTER 4
the molecular network

1. Conditions for rubber-like elasticity

IN this chapter we shall follow up the question raised at the end of the last chapter and consider in greater detail the conditions which have to be satisfied if a polymer is to exhibit the property of rubber-like elasticity. This consideration will require an examination not only of the properties of the polymer molecules themselves, but also of the way in which these molecules are arranged in relation to one another and of the forces by which they are held together. From such considerations we shall be led to a much more precise understanding of the properties of rubbers, and will also be in a better position to examine the behaviour of polymers in other states, i.e. the glassy, crystalline and molten or viscous states.

There are, in all, three conditions which must be satisfied if a material is to show rubber-like properties. These are

1. It must be composed of long-chain molecules possessing freely rotating links.
2. The forces between the molecules must be weak, as in a liquid.
3. The molecules must be joined together or ' cross-linked ' at certain points along their length.

The first of these conditions has already been fully discussed and need not detain us further. The second is essential if the molecules are to have the freedom of motion and the ability to change their conformations in accordance with the basic postulates of the kinetic theory of rubber elasticity. To understand what is meant by the forces being *weak* we must consider for a moment the difference between a liquid and an ordinary hard solid, such as a crystal or a glass. As explained in the preceding chapter, the molecules in the normal solid are bound together by relatively strong forces to form a rigid geometrical structure in which each molecule occupies a definite position with respect to the molecules immediately surrounding it (cf. fig. 3.1). The thermal motion of the molecule takes the form of an oscillation about a fixed mean position, but the amplitude of this oscillation is not great enough to enable any given molecule to break away from the fields of force by which it is bound to its neighbours. In a liquid, on the other hand, the inter-molecular forces are not so strong, and individual molecules do have sufficient kinetic energy to break away from their immediate neighbours

44

and form other attachments. It is this repeated breaking away of individual molecules which gives the liquid its fluidity; the structure is not rigid, but is characterized by a continued breaking down of existing groupings of molecules and re-formation of new groupings.

The meaning of the condition referred to in (2) above is that for a material to be rubbery it must have this same kind of loose structure which enables the molecules, or segments of molecules, to rearrange themselves in various ways in response to their thermal agitation. The materials of which rubbers are composed are of such a chemical nature that the forces between their molecules are similar to the forces between the molecules of an ordinary liquid, like paraffin, for example.

We now come to the real difficulty. If the forces between the molecules of a rubber are sufficiently weak for them to be able to move freely with respect to one another, like the molecules of a liquid, why should the material in fact be a solid rather than a liquid?

The difficulty may be overcome by postulating that the long-chain molecules of the polymer shall be joined together at certain points along their length. Owing to the great length of the chains quite a small number of such ' cross-linkages ' is sufficient to ensure that each molecule is connected to at least two other molecules so that the whole assembly of chains becomes effectively a single structure, as shown in fig. 4.1 (a).

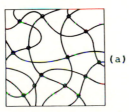
(a)

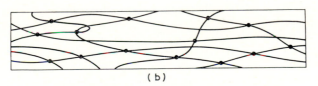
(b)

Fig. 4.1.   Cross-linked network (a) in unstrained state and (b) in strained state.

These cross-linkages do not interfere appreciably with the local freedom of motion of the segments of molecules with respect to segments of neighbouring chains; they do not, therefore, interfere with the mechanism on which the phenomenon of rubber-like elasticity rests; but they do eliminate the possibility of bulk slippage of one molecule past another, which is the necessary condition for flow to take place. The combination of the second and third of our three basic conditions thus enables us to satisfy the two opposing requirements, the requirement of freedom of local molecular motion and the requirement of suppression of flow, and hence to combine in a single material certain of the properties of both the liquid and the solid states.

The process of introducing the cross-linkages into the rubber is known as *vulcanization*; this process involves a chemical reaction with sulphur. Vulcanization leads to a great improvement in the elastic recovery of the rubber and practically eliminates creep or flow effects. Before we consider this in detail, however, it is desirable to examine another possible means whereby the third of our fundamental conditions for rubber-like elasticity may be satisfied.

2. *Chain entanglements*

Natural rubber in the raw or unvulcanized state, though less perfectly elastic and more subject to irreversible effects than vulcanized rubber, does nevertheless exhibit the characteristic long-range extensibility of the vulcanized material. Its response to short-time loading is not very different from that of vulcanized rubber; an unvulcanized rubber ball, for example, shows a normal rebound resilience or ' bounce '. It is only for longer periods of loading, particularly under high stresses, that the superiority of the vulcanized material is apparent. It must therefore be accepted as a fact that the presence of cross-links in the chemical sense is not an absolute necessity for the existence of a considerable degree of rubber-like elasticity. How is this apparent inconsistency to be reconciled?

The answer to this question is that the function of the cross-linkages may be taken over by physical entanglements between the molecules. Some degree of entanglement is a necessary consequence of the randomly kinked conformation of the very long chains. The forces between these chains, though weaker than those in a rigid solid, are by no means negligible, and the effect of complex local entanglements is to magnify their effect, just as a knot or entanglement in a skein of wool magnifies the normal amount of friction between adjacent lengths of the yarn. Such entanglements therefore produce regions in which the resistance to relative motion of the molecules is very much above the average, and these regions, though more spread out than the points of chemical cross-linkage in the *vulcanized* rubber, act in a very similar manner, at least if the time-scale considered is relatively short. For longer times of loading, however, the individual entanglements slowly shake themselves out, and this results in appreciable flow or creep.

3. *Charles Goodyear and the discovery of vulcanization*

Charles Goodyear, in America, was much concerned with the defects of natural rubber in the early days of the rubber industry. The most serious of these was the inherent tendency to flow, noted above, which led to permanent loss of shape or distortion under load, as well as to an unpleasant surface stickiness which was particularly objectionable in footwear or other articles of clothing. It was because of this stickiness that the first rubberized cloth, produced by Macintosh[†] in Glasgow in 1823, was made by sandwiching a thin layer of rubber between two sheets of fabric. Another trouble was the stiffening or loss of elasticity which occurs at low temperatures, which we now know to be due to crystallization. Goodyear spent many years of his life in efforts to overcome these defects, particularly the creep and surface stickiness. Success eventually came in the year 1839 with the discovery of vulcanization. The story of this discovery is described by Goodyear himself in his book *Gum-Elastic*, published in 1855—a fascinating document of equal interest for the revelations it gives of the character of the man as of the state of the rubber industry at the time. Goodyear had become interested in the effect of incorporating sulphur into rubber, which had been found by Hayward to confer some benefit. He discovered that the effect was enhanced by exposure to sunlight. In marked contrast to the behaviour of pure rubber, sheets of rubber into which powdered sulphur had been mixed completely lost their surface stickiness when exposed to sunlight. Attempts to exploit this advantage commercially, however, were unsuccessful, chiefly because the improvement was limited to a very thin surface layer; the properties of the bulk of the material were unchanged. Goodyear then went on to examine the effect of heat on rubber containing sulphur, and it was in the course of these experiments that the critical discovery was made. He writes: ' He [the inventor] was surprised to find that the specimen, being carelessly brought into contact with a hot stove, charred like leather. He endeavoured to call the attention of his brother, as well as some other individuals who were present, and who were acquainted with the manufacture of gum-elastic, to this effect, as remarkable, and unlike any before known, since gum-elastic always melted when exposed to a high degree of heat.'[‡]

Goodyear followed up this hint—and it was no more than a hint—of a transformation in properties by systematic investigations, and was soon able to demonstrate that on less severe heating a product was obtained which had greatly improved elastic properties and did not become sticky at high temperatures.

The discovery of vulcanization, as Goodyear himself acknowledged, owed something to accident. To attribute it entirely to accident, however, would be to overlook the long years of preparation and experiment which preceded it. Without the familiarity with the material born of years of experience, combined with a clear vision of the ultimate objective, it is most unlikely that Goodyear would have given a second thought to his original casual observation. The truth is fairly expressed

[†] The *k* in the word *mackintosh* is an intrusion.
[‡] Charles Goodyear, *Gum-Elastic*, Vol. I. New Haven, 1855, p. 118.

in his own words: '. . . it has sometimes been asked how the inventor came to make the discovery. The answer has already been given. It may be added, that he was many years seeking to accomplish this object, and that he allowed nothing to escape his notice that related to the subject. Like the falling of an apple, it was suggestive of an important fact to one whose mind was previously prepared to draw an inference from any occurrence which might favour the object of his research. While the inventor admits that these discoveries were not the result of *scientific* chemical investigations, he is not willing to admit that they were the result of what is commonly termed accident; he claims them to be the result of the closest observation and application.'†

Not only did the process of vulcanization as discovered by Goodyear succeed in eliminating the effects of flow, i.e. surface stickiness and creep under load, it also, as a kind of bonus, substantially eliminated the progressive hardening which takes place in raw rubber at temperatures around $0°C$ or below, which is now known to be due to crystallization (cf. Chapter 6). Goodyear clearly appreciated the enormous industrial significance of his discovery, and was impatient with the time which it took for its value to be recognized. Meanwhile, he existed in a state of acute poverty and suffered imprisonment for debt. Goodyear's process has been used practically unchanged in the manufacture of articles of all kinds from the time of its discovery until the present day; without it, the rubber industry as we know it would hardly have been possible.

4. *Deformation characteristics of rubber*

From the ideas concerning the structure of a vulcanized rubber which have been discussed in the preceding pages we are able to form a concept of a kind of ideal rubber, similar in principle to the concept of an ideal gas as developed from the kinetic theory of gases. The ideal rubber is visualized as a loose 3-dimensional network of chains, of irregular or random form, joined together by unbreakable chemical bonds or cross-linkages. Except for these points of cross-linkage the forces between the chains are assumed to be negligible; each chain is assumed to be free to take up any desired conformation.

We come now to the question of working out the properties of an ideal rubber of the type envisaged. In order to understand what is involved in this problem we must first take a look at some of the mechanical properties of an actual rubber in more detail. Let us suppose that we take a strip of vulcanized rubber, such as an ordinary rubber band, fix one end, and apply a load‡ to the other end. By varying the load and measuring the length corresponding to each value of the load we obtain the force-

† Charles Goodyear, Gum-Elastic, Vol. I. New Haven, 1855, p. 120.
‡ Since this is what is measured, we shall speak of it as a single force. It must be remembered, however, that the state of stress in the specimen is the result of equal and opposite forces, one at each end.

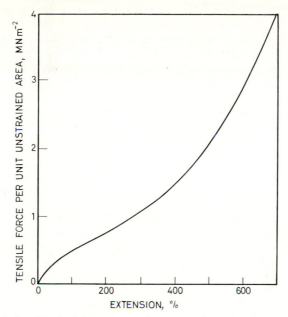

Fig. 4.2. Typical force–extension curve for vulcanized rubber.

extension curve for the rubber. A typical result for natural rubber is shown in fig. 4.2, in which the force per unit *unstrained* cross-sectional area is plotted against the percentage extension.

The most obvious feature of this result is that the relation between the applied force and the resultant deformation or *strain* is not a linear one; the force is *not* proportional to the extension. This behaviour is in marked contrast to the behaviour of ordinary solids. For such materials the law of elasticity originally propounded by Hooke, namely that in any elastic deformation the *stress* is proportional to the *strain*, is satisfied. Rubber evidently does not obey Hooke's law when subjected to a tensile stress.

Our examination of the mechanical properties of rubber, however, need not be limited to a simple extension. One of the most interesting aspects of the subject is concerned with the different types of behaviour observed under different kinds of deformation. A cylindrical rod of rubber, for example, may be subjected to twisting or torsion or (if it is not too long) to compression along its axis or shearing parallel to its base plane as well as to extension. For any given type of strain there is a characteristic form of stress–strain relation, and any comprehensive treatment of the mechanical properties of rubber must take into account these different types of strain.

The deviation from Hooke's law shown in fig. 4.2 is associated with the fact that in the case of rubber we are concerned with deformations which are very large. In ordinary elastic solids the deformations to be

49

considered are very small, that is to say of the order of about 1%. The classical theory of elasticity, as normally applied in physics and engineering, is a theory of *small* elastic deformations, and it is only in the region of small deformations that (in general) Hooke's law is applicable. (Hooke's law in fact applies to a rubber, provided that the strain is limited to 1%.) When we come to large elastic deformations, of the kind obtainable only in rubber-like materials, a very different approach to the problem of defining the elastic properties has to be evolved. It is this new form of approach which the molecular theory of rubber elasticity has successfully provided.

5. *Geometry of strain*

Before proceeding further, attention must be drawn to one important basic property of rubbers which greatly simplifies the mathematical treatment of their elastic properties. This is that all the different types of deformation that may be applied take place without any appreciable change of *volume*. The effect of this can be illustrated by considering again a simple extension. Suppose we start with a block of rubber in the form of a cube of edge length l_0 and apply an extension in the direction of one of the edges such that this edge increases to the length l_x (fig. 4.3).

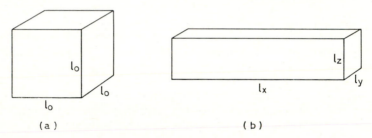

Fig. 4.3. Deformation at constant volume. (*a*) Unstrained state. (*b*) Strained state.

In the classical theory of elasticity it is usual to define the strain as the increase of length divided by the original length, i.e.

$$\text{Strain} = \frac{l_x - l_0}{l_0}. \tag{4.1}$$

In large-deformation theory it is more convenient to define the strain in terms of the *extension ratio*, which will be denoted by λ (lambda). Thus

$$\text{Extension ratio} = l_x/l_0 = \lambda. \tag{4.2}$$

With this definition, the change in the lateral or transverse dimensions, l_y and l_z, obtained from the condition for constancy of volume, takes a very simple form. The volume in the unstrained state is l_0^3, while in the strained state it is $l_x l_y l_z$. Hence

$$l_x l_y l_z = l_0^3. \tag{4.3}$$

50

Since the two lateral directions are equivalent it is obvious that $l_y = l_z$, while from (4.2) $l_x = \lambda l_0$. Inserting these values into equation (4.3) we obtain

$$(\lambda l_0)l_y{}^2 = l_0{}^3.$$

Hence

$$l_y = l_z = \frac{1}{\sqrt{\lambda}}\, l_0. \tag{4.4}$$

The result (4.4) means that the two lateral dimensions are reduced in the ratio $1/\sqrt{\lambda}$. The shape of the specimen in the deformed state is therefore completely determined by the specification of the single strain parameter λ.

This simple *strain geometry*, as it may be called, is not observed with ordinary solids, such as, for example, steel or glass. If a rod of steel is extended, the lateral dimensions contract, but the amount of this contraction cannot be calculated on general grounds; it has to be determined by experiment. The ratio of the lateral contraction to the longitudinal extension, which is called Poisson's ratio, has a particular value for each material. The meaning of this is that for ordinary hard solids there is an *increase* in volume on extension, and the amount of this increase (which is directly related to the value of Poisson's ratio) varies from one material to another. In these materials, therefore, it is not possible to define the complete state of strain in terms of a single parameter; we need to know not only the amount of the extension but also the value of Poisson's ratio.

6. *Significance of constancy of volume*

This peculiar property of rubber, namely the maintenance of constancy of volume during deformation, is a direct result of its peculiar mechanism of elastic deformation. We have seen that this deformation arises from the changes in conformation of the molecules in an open network structure of the type represented in fig. 4.1. The forces required to produce such a network deformation are comparatively weak; this is why the modulus of elasticity of rubber (as was noted in the preceding chapter) is so very much lower than that of a material such as steel. The *volume* of a rubber, however, is determined by the actual volume occupied by the molecules themselves, and this is in no way related either to the *conformations* of the molecules or to the existence of a cross-linked network; this is borne out by the observation that the process of vulcanization has no appreciable effect on the density of the rubber. The volume of the material is maintained by the forces *between* the molecules, just as in the case of any other material. The deformation of the *network* is therefore a quite separate process, which can be carried out without affecting the lateral forces between chains, and hence without affecting the volume.

By contrast, if we recall the model of the structure of a solid (Chapter 3, § 1), we can see that the same forces which determine the separation

51

between neighbouring atoms also determine the elastic response to a stress; the change of volume is therefore of the same order of magnitude as the longitudinal extension.

A quite different situation exists if we consider the response of rubber to an applied *hydrostatic pressure*. The effect of a hydrostatic pressure is to reduce the volume; this corresponds to a uniform reduction in all three dimensions. This reduction of volume is brought about by squeezing the molecules closer together, and is defined by the volume compressibility. This genuine compressibility is directly related to the forces *between* the molecules, which in the case of a rubber are very similar to the forces between the molecules of a typical liquid; it is essentially unrelated to the network properties which are involved in a tensile deformation. The compressibility of rubber is actually almost identical with that of water. The important point is that the changes in dimensions produced by a hydrostatic pressure are extremely small compared with the changes produced by a tensile stress of similar magnitude. For example, a hydrostatic pressure of 10^6 N m^{-2} (10 atmospheres) would reduce the volume of a rubber by only 5 parts in 10 000, but a tensile stress of 10^6 N m^{-2} would produce an increase of length of at least 100%. Hence for practical purposes rubber may be regarded as substantially *incompressible*, and changes of volume in extension or other types of deformation may be regarded as completely negligible.

7. Properties of a molecular network

The above discussion has taken us rather a long way from our starting point, which was to consider the properties of an idealized molecular network. Let us take a network, such as that depicted in fig. 4.1 (*a*), containing a specified number of cross-linkages, and hence a specified number of 'chains', a chain being defined, in this context, as the segment of a molecule between successive points of cross-linkage. In the unstrained state this network may be taken to be a cube of unit edge length. On application of a tensile force F the network will be distorted, as shown in (*b*), the changes in longitudinal and lateral dimensions being related in the manner considered above. The problem is to calculate the relation between the force F and the corresponding extension ratio λ.

This problem has been attacked in a number of different ways, all of which lead to the same conclusion. It would take us too far to go into any one of these methods in detail. There is, however, one rather simple way of looking at the problem which will serve to illustrate the kind of reasoning involved. We saw in Chapter 3 that the single molecule behaves like a little spring which obeys Hooke's law (stress proportional to strain). The network of molecules may thus be regarded as a network of springs joined together at their ends. An element of this network may be represented by a single junction point (or cross-linkage) O (fig. 4.4 (*a*), from which four springs radiate outwards to other junction

52

points at A, B, C and D. Assuming for the moment that these outer junction points are fixed, the position of the central junction point will be determined by the requirement that the set of forces acting on the point O shall be in equilibrium.

Now let us assume that the outer junction points are displaced to new positions A′, B′, C′, D′ corresponding to the strained state (fig. 4.4 (b)).

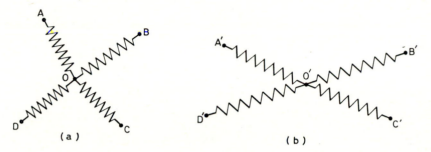

Fig. 4.4.   Representation of element of network in (a) unstrained and (b) strained state.

The central junction point will now move to some new position O′ at which the forces in the extended chains are again brought into equilibrium. For the type of spring considered there is a very simple relation between this new position and the original one. This is found to be given by the condition that *the changes in length and direction of the lines OA, OB, OC and OD correspond exactly to the changes in length and direction of a corresponding set of lines drawn on the bulk rubber.*

From this reasoning it follows, by a similar argument, that if we fix the positions of each of the junction points in the unstrained state of the network we can calculate their corresponding positions in the strained state. We are thus able to relate the lengths of the chains in the strained state to their lengths in the unstrained network. From this it is not very difficult to calculate the forces exerted by the whole system of chains on the boundary surfaces, and hence to derive the total force F acting on the specimen.

It turns out that the result of this calculation is completely independent of the particular choice of junction-point positions in the unstrained state. The result is therefore quite general, and independent of the detailed structure of any particular network, which of course cannot be precisely specified, since the cross-linkages are introduced in a random manner. The theoretical relation between the force F (per unit unstrained cross-sectional area) and the extension ratio λ is expressed by the equation

$$F = G(\lambda - 1/\lambda^2), \qquad (4.5)$$

where G is a constant.

53

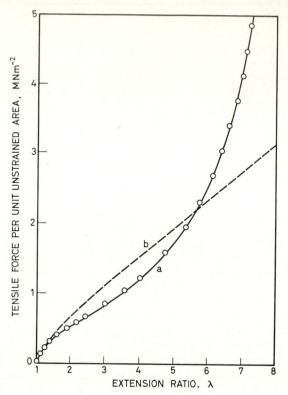

Fig. 4.5. Force–extension curve for cross-linked rubber.
(a) Experimental. (b) Theoretical.

The form of this relation is shown in fig. 4.5 together with a typical experimental curve for vulcanized rubber. The theory leads to the important conclusion that the ideal rubber should not obey Hooke's law, and it succeeds in reproducing approximately the observed form of force–extension curve up to an extension of about 400% ($\lambda = 5$). At still higher extensions special considerations (associated with the limited extensibility of the chains) arise, and the theory in the form presented ceases to be applicable.

Simple shear

If equal and opposite forces are applied *tangentially* to a rectangular block, the type of deformation produced is called a *simple shear* (fig. 4.6). In this type of strain planes parallel to the base plane are displaced in such a way that the rectangle ABCD is transformed to the parallelogram, A′B′CD, the vertical edge being tilted through the angle ϕ. The shear strain γ is measured by the *tangent* of the angle ϕ, i.e.

$$\gamma = \tan \phi. \qquad (4.6)$$

54

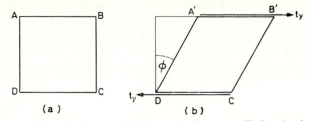

Fig. 4.6.   Simple shear.   (*a*) Unstrained state.   (*b*) Strained state.

The statistical theory may be applied just as easily to the problem of shear as to simple extension.   In this case the resulting relation between the shear stress t_γ (i.e. the tangential force per unit area) and the shear strain is given by

$$t_\gamma = G\gamma, \qquad (4.7)$$

where G is the same constant as in equation (4.5) for simple extension.

This very simple result means that the shear stress is proportional to the shear strain.   We thus have the remarkable result that *a rubber should obey Hooke's law in shear, but not in extension.*

This conclusion is also confirmed by experiment, at least approximately (fig. 4.7).   Up to a shear strain of about 1·0 ($\phi = 45°$), the agreement is very good; beyond this the experimental curve falls somewhat below the theoretical line.

Even more significant than the general agreement in form of the experimental and theoretical curves for simple extension and shear is the agreement in the value of the elastic constant G required to fit the experimental data for these two types of strain.   The experiments referred to were carried out on the same sample of vulcanized rubber, and the theoretical curves in figs. 4.5 and 4.7 were calculated using the *same* value of this constant.

Other relations may be worked out for other types of strain, e.g. axial compression of a cylinder, 2-dimensional extension of a sheet, etc., and in all cases the degree of agreement between theory and experiment is comparable with that shown in the particular cases of extension and shear considered above.   Two features of these results are of particular interest, i.e.,

(1) simple shear is the only type of deformation in which Hooke's law is obeyed; in all other types of deformation the stress is a non-linear function of the strain;

(2) the properties of a rubber in any type of strain may be defined in terms of a single elastic constant G.

8. *Numerical value of the modulus*

There is still one aspect of the network theory to be examined.   This is concerned with the numerical value of the elastic constant G.   The

55

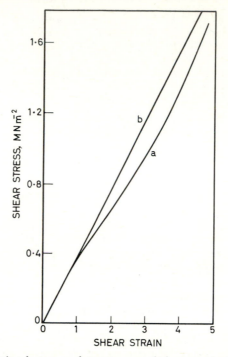

ig. 4.7.   Relation between shear stress and shear strain for cross-linked
rubber.   (*a*) Experimental   (*b*) Theoretical.

theory shows that this constant, which, as will be seen from equation
(4.7) is equivalent to the shear modulus, is determined by the number N
of ' chains ' per unit volume of the network, a ' chain ' being defined as
the segment of molecule between the successive points of cross-linkage.
The theoretical relation is

$$G = NkT, \qquad (4.8)$$

where T is the absolute temperature and k is Boltzmann's constant.
The appearance of T in this expression represents the fact, already fully
examined in earlier chapters, that for any type of strain the stress is
directly proportional to the absolute temperature in Kelvin.

The value of N, the number of chains per unit volume, is determined
by the number of cross-linkages introduced in the vulcanization process;
the greater the number of cross-linkages, the greater will be the number
of chains.   Closer examination shows that each new cross-link introduced
leads to the formation of two new chains, hence N is equal to twice the
number of cross-links per unit volume.

It follows from this that if we could introduce a measurable number of
cross-links, and hence derive the value of N, it should be possible to

56

calculate the value of the elastic constant G independently of any measurement of the stress. A comparison of the value calculated in this way with the value obtained directly from the observed force–extension curve (e.g. in simple extension) should thus provide a further check on the *quantitative* validity of the theory.

On account of its fundamental significance a great deal of attention has been devoted to this problem. The chief difficulty, experimentally, is to find a method of cross-linking in which there is a definite relation between the number of cross-linkages produced and the amount of the chemical reaction between the cross-linking agent and the rubber. This has involved a close study of a number of cross-linking reactions, using low molecular-weight materials in which the products of reaction can be isolated and analysed. Sulphur has been found to be unsuitable for the purpose; it produces various proportions of cross-linkages containing two, three or more sulphur atoms as well as monosulphide linkages. A reagent which satisfies the requirements in di-tertiary butyl peroxide. We need not concern ourselves with the chemistry of the process, except to note that it produces direct C–C cross-linkages between chains, of the type

$$-CH-C(CH_3)=CH-CH_2-$$
$$|$$
$$-CH-C(CH_3)=CH-CH_2-$$

the reactant itself disappearing from the final product. The number of cross-linkages formed is directly related to the amount of the reagent consumed; this is obtained by chemical analysis.

The work in which this cross-linking agent was utilized was carried out by Moore and Watson in 1956, and represents the most careful and advanced study of this problem yet undertaken. The results are represented in fig. 4.8. The horizontal scale gives the value of the modulus G obtained from the chemical estimate of the amount of cross-linking (equation (4.8)), while the vertical scale gives the corresponding directly measured value of the modulus. If the theory applied exactly the result would be represented by a straight line at an angle of 45° passing through the origin. The figure shows that the slope is nearly correct, but that the experimental line is displaced upwards from the theoretical line. This suggests that each additional cross-link produces the expected addition to the modulus, but that there are a certain number of ' physical ' cross-linkages initially present in the rubber, which impart to it a certain modulus even in the absence of chemical cross-links. This interpretation is in harmony with the argument already put forward earlier in this chapter to account for the observation that even in the unvulcanized state rubber possesses a definite degree of elasticity. The ' physical ' cross-linkages postulated by Moore and Watson may be identified with the molecular entanglements previously discussed.

The deviation from the predicted value of the modulus is most serious in the early stages of cross-linking. Over the range encountered in

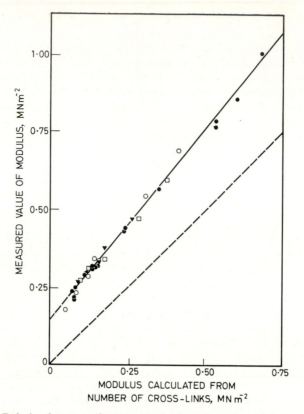

Fig. 4.8. Relation between the measured value of shear modulus and the value calculated from the chemically estimated number of cross-linkages. If the theory applied exactly these would be identical, as represented by the dashed line. (Adapted from Moore and Watson (1956).)

typical vulcanized rubbers, the deviation from the theory is not more than about 25%. In considering the significance of this result, it should be borne in mind that no adjustable parameters of any kind are introduced into the theoretical calculation. Absolute numerical predictions of this kind are very rare in physics, and the prediction of the absolute value of the modulus to this degree of accuracy on the basis of a very general mathematical model must be regarded as a striking success for the statistical theory.

9. *Degradation and network breakdown*

Before leaving this subject of network properties some mention should be made of the processes of chemical degradation and network breakdown. These processes lead to loss of elasticity, surface cracking and other deleterious effects, known generally as ' ageing ' or ' perishing '.

Natural rubber is especially prone to these degradative reactions, which are associated with the presence of the more highly reactive C=C double bond in the structure of the isoprene unit (formula (1.7), p. 3).

This enhanced reactivity, however, may also be beneficial. It is responsible for the ease with which the vulcanization reaction may be carried out. Polymers which do not contain a double bond cannot easily be cross-linked, and in the case of polyisobutylene, for example (see table, Chapter 5), it is necessary to incorporate a small proportion of a double-bonded monomer such as isoprene to obtain a vulcanizable rubber (butyl rubber). Again, in processing rubber, it is necessary to carry out a preliminary milling operation, during which the very long rubber molecules are broken down by the combined effects of mechanical stress (shearing), high temperature and atmospheric oxygen. It is only after such a breakdown that the material becomes sufficiently soft or 'plastic' for the vulcanizing and other compounding ingredients (e.g. carbon black, pigments, etc.) to be mixed in, and for the end product to be formed by rolling or moulding. These shortened molecules are then joined together again during vulcanization to form the permanent network, thus fixing the final form of the required product.

The normal process of ageing is due to various types of chemical reaction with oxygen, particularly with oxygen in the form of ozone (O_3). These reactions are generally accelerated by heat and also by light. The reaction may lead to breakdown of either the chains or the cross-linkages in the vulcanized rubber, with consequent disruption of the network structure and loss of elastic properties and strength. The presence of mechanical strain, which leads to the opening up of incipient surface cracks and minute fissures, and consequent exposure of fresh surfaces to chemical attack, has been shown to intensify the rate of degradation. Hence, to obtain the longest useful life rubber should, wherever possible, be kept in a cool place, away from direct sunlight and preferably not subjected to strain.

On account of its sensitivity to breakdown natural rubber has been replaced by other materials for most purposes where long life is a primary consideration (e.g. in electrical wiring) or where high temperatures are encountered. Better performance under such conditions is obtained from polychloroprene rubber or polyvinyl chloride, both of which have greater chemical stability than natural rubber. This does not apply, however, to tyres, for which the outstanding requirement is high strength and resistance to abrasion and wear, for which natural rubber is superior.

10. Conclusion

In this chapter we have pursued the story, begun in the preceding chapter, of the discovery of the nature of rubber and the mechanism of rubber elasticity. The properties which at first appeared so mysterious, and which for more than a century had defied all attempts at a rational explanation, are now seen as a logical consequence of the molecular structure of the material. These properties are as well-defined, and as capable of quantitative mathematical description, as are the properties of other classes of materials, such as, for example, gases or crystalline solids. Natural rubber is no longer unique: its properties may be approached— if not exactly reproduced—by a whole range of synthetically produced polymers in everyday use. Indeed, far from being regarded as complicated and difficult to understand, rubbers today are more usually thought of as rather simple materials, having rather well-defined and easily understood properties. This, of course, is by comparison with other polymeric materials, more particularly crystalline polymers and fibres, which, as we shall see in later chapters, present problems of description and interpretation which are still far from being fully resolved.

At this stage it is interesting to recall some of the earlier theories of rubber elasticity, referred to in the last chapter. These, it will be remembered, were mainly based on two distinct principles: (1) the coil spring molecule and (2) the open network. These two principles have in fact survived together—though in a somewhat modified form—in our present-day concept of rubber elasticity. The molecule is indeed a kind of spring, though a spring of a very different kind from an ordinary coil spring, and the structure of the rubber is like a network, but a network which is also of a rather peculiar kind, comprising the whole assemblage of molecules, not a separate component of a two-phase structure. The important distinction between our present interpretation and these earlier theories lies not in these *general* ideas, but in the *specific* mechanism to which the elasticity is attributed. It was the kinetic or *statistical* concept which provided the key to the problem, and which created a revolution in thought in the field of polymers comparable to the revolution in thought in the wider field of physics and chemistry brought about by the introduction of the atomic theory of matter by John Dalton in the earlier part of the nineteenth century, which, among other things, gave the first satisfactory explanation of the laws of gases established experimentally some two centuries earlier.

60

1. *What is a glass?*

GLASSES, like rubbers, are of great scientific interest, but for reasons which are almost diametrically opposite. The characteristics of rubbers are flexibility and toughness, high deformability and the capacity to withstand impact. The characteristics of glasses are extreme rigidity combined with brittleness or fragility. Glasses and rubbers stand at opposite extremes in the scale of mechanical properties.

Glasses are also like rubbers in being a rather peculiar form of matter. They may on occasions exhibit spontaneous fracture without warning and for no immediately obvious reason*, though this, fortunately, is a rather rare occurrence, and may be avoided by a careful annealing treatment. We are all familiar, however, with the glass which cracks when dipped into hot water, thus demonstrating its sensitivity to thermal shock, and the spectacle of the car windscreen instantly rendered opaque by thousands of cracks when struck by a flying stone is all too common. The other materials with which we come into daily contact usually give us some warning when things begin to go wrong; they are not in the habit of demonstrating their physical properties with quite such dramatic effect.

It is easier to define a glass in terms of what it is *not* than in terms of what it *is*. Whatever the definition, it must be a definition in terms of *structure*, not in terms of chemical constitution, and only to a limited extent in terms of physical properties. Briefly, a glass is a hard solid which is *not crystalline*. The structure of a glass is not regular, like that of a crystal, but *amorphous*, like that of a liquid. A glass is therefore a solid having a geometrical structure which is characteristic of the liquid state. A glass is formed by the solidification of a liquid without a change of state such as that which occurs in the freezing of a liquid to the more usual crystalline state. In order to understand the glassy state it is therefore necessary to take a brief look at the crystalline state, with which it is to be contrasted. We shall also have to try to find an answer to the question of why it is that, while most materials crystallize on solidification, the glasses prefer to retain the liquid type of structure when solidified by cooling.

---

* The author once tried to convince a lady that her little boy was not necessarily untruthful when he denied having touched a glass which was found broken.

2. Crystals and glasses

Our knowledge of the structure of materials, and particularly of the detailed arrangement of the atoms in a crystalline solid, is derived primarily and most directly from X-ray studies. X-rays are able to pass through matter, but in so doing they are to some extent scattered by the individual atoms. The pattern of this scattering, or *diffraction pattern* as it is called, may be recorded on a photographic plate and analysed to give information about the relative geometrical disposition of the scattering centres, i.e. the atoms, in the structure. The pioneers in this type of investigation were the Braggs, the late Sir William Bragg and his son Sir Lawrence, who together founded the science of X-ray crystallography.

X-ray diffraction effects depend on the phenomenon of *interference* between trains of waves. The principle may be illustrated by the diffraction of a beam of light by an optical *diffraction grating*, which consists of a set of parallel lines ruled on a reflecting plate or mirror at a separation comparable with the wavelength of light (fig. 5.1). The wavelets from individual scattering points arriving at the plane surface BC normal to the wave-front of the ' diffracted ' beam will be *in phase* and so reinforce one another completely only if a certain relation exists between the angle ϕ of the beam and the spacing d of the grating, such that the

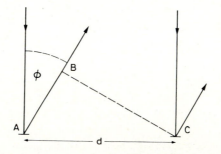

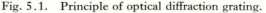

Fig. 5.1. Principle of optical diffraction grating.

62

path difference AB between successive wavelets is an exact number of wavelengths. This relation is

$$d \sin \phi = n\lambda, \qquad (5.1)$$

where λ is the wavelength and n may take the integral values 1, 2, 3… etc. The diffraction pattern consists of a series of lines representing angles ϕ_1, ϕ_2, ϕ_3, … which satisfy the above equation. For the pattern to be obtained it is necessary that the wavelength of the waves shall be of the same order of magnitude as, but less than, the spacing of the lines. Diffraction effects in crystals cannot be observed with light, because the spacing of the scattering centres or atoms is less than one thousandth of the wavelength of visible light. But X-rays have wavelengths of about 10^{-10} m (1 ångström), which is comparable with the separation of atoms, and hence may be used in a similar manner. A slight difference is introduced by the fact that we are now dealing with an array of atoms in three dimensions, and we are concerned with scattering from *successive layers* of atoms in depth separated by the distance d (fig. 5.2). For this

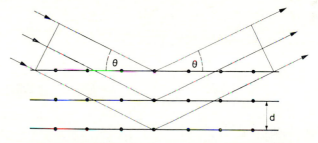

Fig. 5.2.   Diffraction of X-rays by crystal lattice.

case reinforcement occurs only when the angle of reflection is equal to the angle of incidence and when the angle θ at which the beam strikes the surface satisfies the relation

$$2d \sin \theta = n\lambda. \qquad (5.2)$$

This relation is known as Bragg's law.

If a material in the form of a single crystal (e.g. calcite or rocksalt) is inserted in the path of a fine beam of X-rays of wavelength λ, reflected beams will be observed only when the angle of incidence θ satisfies the above relation. To obtain a complete diffraction pattern it is therefore necessary to rotate the crystal and so to pick out the various planes for which Bragg's relation is satisfied. The resulting beams form a pattern of spots on the photographic film like that in Plate 1 (*b*) and (*d*). If the material is in the form of a fine crystalline power this rotation is unnecessary, for there will always be some crystals whose directions happen to be such that Bragg's law is satisfied. The resultant diffraction pattern or

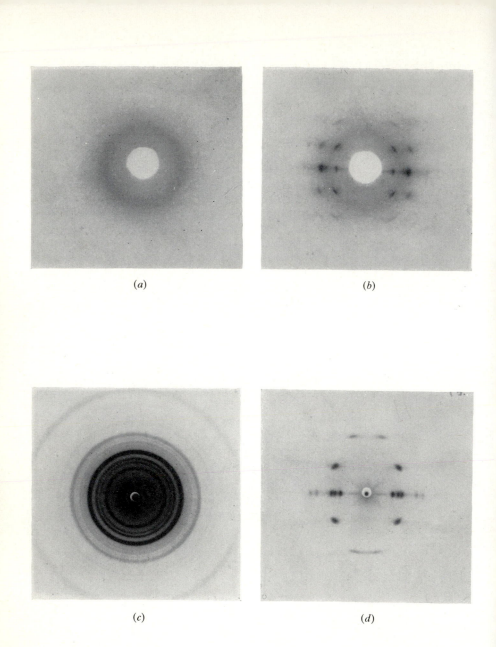

(a) (b)

(c) (d)

PLATE 1. X-ray diffraction photographs. Top: Vulcanized rubber (a) unstrained, (b) stretched to about 650% extension (I. H. Hall). Bottom: Polypropylene (c) unoriented, (d) highly oriented (J. Mann).

64

' powder diagram ' takes the form of a series of complete circles or sharp rings, each ring corresponding to a particular value of θ, that is to a particular pair of values of d and n (Plate 1 (c)).

The sharpness of the rings in the ' powder diagram ' depends upon the size or dimensions of the crystals. As these are reduced below a certain value, the reflected beams begin to spread out, with the result that the rings become slightly broadened. The effect becomes important when the number of parallel planes is less than about 1000, corresponding to crystal dimensions of a few tenths of a micrometre (10^{-6} m), which is somewhat below the limit of visibility in a high-power microscope. As the size is further reduced the rings become more and more diffuse and the whole pattern becomes blurred.

In the liquid state the atoms do not have any definite geometrical regularity of packing as in the crystal. The difference is shown diagrammatically in fig. 5.3. The packing is not *entirely* random, however (as it is in a gas); there is still a certain preferred separation between any one atom and its neighbours, but the actual separation varies considerably between different pairs of atoms. Moreover, the rudimentary regularity

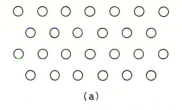

(a)

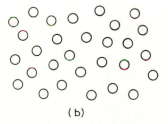

(b)

Fig. 5.3. Diagrammatic representation of arrangement of atoms in (*a*) crystal and (*b*) liquid.

which exists is limited to the immediate neighbourhood of any given atom; as we move out to distances of 4 or 5 atomic diameters, all semblance of regularity is lost. This situation is summed up by the statement that a liquid possesses no *long-range* order. The liquid type of structure may be regarded as the ultimate end-point if the size of the crystallites in a powder (or polycrystalline solid) is reduced indefinitely, i.e. to dimensions corresponding to the distances between individual atoms. In this state there is no definite spacing between planes of

65

atoms—indeed there are no planes of atoms—and the rings in the diffraction pattern give way to a single very diffuse and ill-defined ring or *halo*, as shown in Plate 3 (*a*).

The point to which we are now coming is that the diffraction pattern of a glass is of the same kind as that of a liquid. This is the most direct evidence for our earlier statement that a glass is non-crystalline, and that its geometrical structure is similar to that of a liquid.

We have still to answer the question *why* the atoms or molecules in the glass should prefer to remain in the disordered state, rather than settle down in the seemingly more natural regular state of packing of the crystalline solid. We have already noted that the atoms in the solid are in a state of continual thermal agitation. What is it that prevents them from 'shaking down' into the normal regular arrangement of the crystal?

In considering this question we have to take into account that the unit of structure in the case of a glass is not the single atom, but the molecule. The simple picture represented by fig. 5.3 may be adequate for a monatomic liquid such as a molten metal (e.g. mercury), but such monatomic substances are invariably crystalline in the solid state. The inorganic glass-forming molecules, particularly SiO_2 (silica) and B_2O_3 (boric oxide) have a particular propensity for joining up end-to-end to form irregular 3-dimensional network structures like that represented diagrammatically in fig. 5.4 (*a*). These structures are rather like those found in a typical cross-linked organic polymer (compare fig. 1.2) but the degree of cross-linking is very much higher and the chains correspondingly shorter. Random networks of this kind are essentially non-crystalline: their basic irregularity is incompatible with the formation of a regular crystalline lattice. The corresponding type of crystalline structure for a material of the same chemical composition is represented diagrammatically in fig. 5.4 (*b*).

3. *The process of glass formation*

The formation of a glass is the result of the tendency of the molecules to associate or combine chemically while the material is still in the liquid state. The resulting chain-like and highly branched structures (fig. 5.4 (*c*)) becomes more complex as the temperature is slowly reduced, and these structures greatly increase the viscosity of the melt (cf. p.172). As a result of this increase in viscosity the mobility of the local groups and their capacity for rearrangement into the ideal crystalline lattice is impeded, a point being reached eventually where the structure becomes effectively immobilized in the irregular form. Glass formation is an attribute of rather large molecules characterized by strong intermolecular forces; the very bulk of the associated groups makes it difficult to fit them together in a regular manner. The sugars, which are also of this type, frequently form glasses, and it is often quite difficult to induce

66

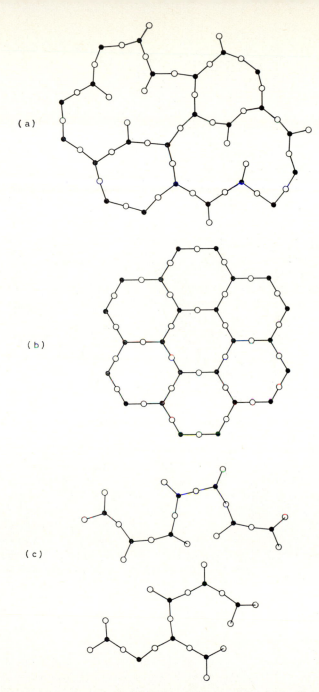

(a)

(b)

(c)

Fig. 5.4. Two-dimensional representation of structure of (*a*) silica glass and (*b*) crystalline silica (crystobalite), (*c*) intermediate linear and branched structures. ● : Silicon atoms, ○ : oxygen atoms.

them to crystallize. The sweets known as ' boiled sugars ', barley sugar, toffees, etc., are essentially glasses. However, the glassy (or vitreous) state (as it is sometimes called) is inherently less stable than the more normal crystalline state, and if the appropriate conditions are not maintained during the glass-forming process crystallization or *devitrification* may make its unwanted appearance.

4. *Glassy polymers*

Whereas in the case of an inorganic glass-forming material we are concerned with the transition from the *liquid* to the solid state, in the case of an organic polymer, the corresponding transition is from the *rubbery* to the glassy state. Except for this difference, the transition to the glassy state in a polymer is entirely similar to that in an inorganic glass. The presence of long chains, and the correspondingly high internal viscosity, is a factor which automatically tends to favour the irregular glassy state rather than the regular, crystalline state. Nevertheless many polymers do crystallize on cooling below a certain temperature, and the reasons why some polymers crystallize while others tend to remain in the amorphous or glassy state have still to be considered. This question will be examined in greater detail in the following chapter, when we come to consider the process of crystallization in polymers. Broadly speaking, the factors which inhibit crystallization in inorganic glass-forming materials—bulky molecules and strong intermolecular forces—have a comparable effect in organic or polymeric materials. The chain repeating units for the two most important organic glasses, polymethyl methacrylate (perspex) and polystyrene, are given in the table on p. 72, from which it will be seen that they both contain rather large side-groups attached to the main chain; these side-groups not only give rise to rather strong local intermolecular forces, but also interfere, in a purely geometrical way, with the freedom of motion of the chains.

In addition to these factors, however, there is a further more fundamental condition which has to be satisfied if a polymer is to be capable of crystallization—one which does not arise in the case of non-polymeric materials. This is that the chain itself shall be perfectly regular, that is to say, that the sequence of groups shall be repeated identically along the whole length of the chain. This fundamental condition, as we shall see in the next chapter, is not satisfied by the principal glass-forming polymers.

5. *Nature of the glass transition*

For any particular polymer the change from the rubbery to the glassy state takes place when it is cooled below a certain temperature, called the glass-transition temperature. The nature of the transition may be most simply understood by taking as an example the case of rubber itself. It will be recalled that Meyer, in his examination of the effect of temperature on the elastic tension in stretched rubber, found a sharp discontinuity of slope at a temperature of about $-60°C$ (fig. 3.2). Above this temperature the tension increased with rising temperature, while below this temperature it decreased as the temperature was increased. This discontinuity corresponds to the transition from the characteristic properties of a rubber to the characteristic properties of an ordinary hard solid, e.g. a glass. The rubber used in this experiment was a vulcanized rubber; in the case of raw or unvulcanized rubber the temperature of the transition is somewhat lower, namely $-70°C$.

In passing through the transition temperature rubber completely loses its capacity for large elastic deformations; it becomes hard and brittle. The brittleness may be easily demonstrated by cooling a strip of rubber (or a piece of rubber tubing) in liquid nitrogen and striking it with a hammer. The splinter-type fracture which occurs has all the characteristics of a glassy fracture.

The changes in physical properties on passing through the transition temperature are completely reversible; when the temperature is again raised the rubber immediately reverts to its normal highly elastic condition.

The explanation of these changes in physical properties is quite simple, and follows directly from what we have already learnt about the structure of rubber and the mechanism of rubber elasticity. The basic requirements for rubber-like elasticity, as we saw, were (a) freedom of rotation about bonds within the molecule and (b) weak secondary forces between the molecules. As the temperature is lowered the energy of thermal agitation of chain segments progressively falls until a point is reached at which it is no longer sufficient to overcome the forces between the molecules. When this happens the chain segments become frozen into fixed positions like the molecules in an ordinary solid. Under these conditions the changes of molecular conformation resulting from random rotation about bonds are suppressed and with it the capacity for undergoing deformations of the rubber-like type.

6. *The glass-transition temperature*

The transition from the rubbery to the glassy state is accompanied by changes in other physical properties as well as in hardness and elasticity. One of these is the coefficient of expansion. This may easily be studied by means of a dilatometer (fig. 5.5), in which the polymer is contained in a bulb, with a suitable confining liquid (usually mercury). The glass plug enables the lower end of the bulb to be sealed off without overheating of the specimen. From the rise of the liquid in the capillary, the change of volume on raising the temperature (after correcting for the expansion of the liquid, etc.) may readily be obtained. A typical result is represented in fig. 5.6 for polystyrene; this shows a sharp discontinuity in the rate of expansion at the transition point, which in the case

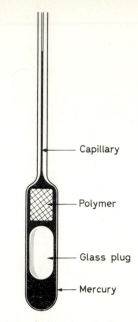

Capillary

Polymer

Glass plug

Mercury

Fig. 5.5.   Dilatometer for study of volume changes.

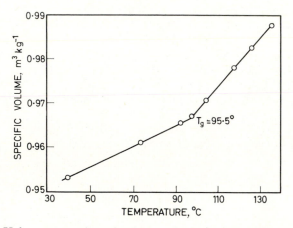

Fig. 5.6.   Volume expansion of polystyrene on heating.   The discontinuity
represents the glass-transition temperature (T_g) (Gordon and Macnab, 1953).

of this particular sample occurred at 95·5°C, the expansion coefficient
above this temperature being about two and a half times its value at lower
temperatures.   It is important, however, to note that there is no change
in the volume itself at the transition temperature, as there would be in the
case of a true change of state, or change of structure, such as occurs on
crystallization.

70

The geometrical structure of a polymer in the glassy state is indistinguishable from the geometrical structure of a rubber. The X-ray diffraction photographs for both states have the form characteristic of a liquid or amorphous structure—a diffuse halo with no sharp rings Pl. 1(*a*). The difference between these two states lies not in their geometrical structure, but in the state of molecular motion. Below the glass transition the molecular segments or groups which form the backbone of the chain can undergo only restricted degrees of vibration and cannot change their positions with respect to segments of neighbouring chains; above the transition complete rotations about bonds, involving relative motion between neighbouring segments, becomes possible. The higher expansion coefficient is a reflection of this greater degree of freedom.

The temperature of the transition is a very important property of a polymer, determining, broadly, whether it is to be classified as a rubber or as a glass. Values of the transition temperature for selected polymers are given in the accompanying table.

Most of the industrially important rubbers have very low transition temperatures, generally below $-50°C$. For the glassy polymers the transition temperatures range from about $80°C$ upwards. Materials whose transition temperatures are not far removed from normal operating temperatures are not generally useful either as rubbers or as glasses, since their properties are too much dependent on the prevailing conditions (see § 7 below).

The rubbery properties of an organic glass above the transition temperature are easily demonstrated. If we take a strip of perspex, for example, which has been heated to $140°C$ in an oven, we find that it is soft and flexible like rubber. It may be easily stretched to several times its length and will recover completely on removal of the stretching force. If, however, it is held in the stretched state and allowed to cool, the deformation will be retained indefinitely, unless the temperature is subsequently raised above the transition point, whereupon it reverts to its original state.

A variation of this experiment is illustrated in fig. 5.7. A strip of perspex (*a*) is heated to $140°C$ and twisted, the twisted form (*b*) being fixed by immersion in cold water. On returning to $140°C$, the original form (*a*) is spontaneously recovered.

7. Mechanical properties in transition region

Though certain of the physical properties, notably the coefficient of expansion, show a sharp discontinuity at the glass-transition temperature, the change in elastic properties is much more gradual and may cover a range of temperature of as much as $50°C$. The region in which this change from the rubbery to the glassy state occurs is called the transition region. The mechanical properties in this region are rather curious; they are very much dependent on the time or rate of application of the

Polymer	Repeating unit	Transition temp. (°C)
Silicone rubber	CH_3 | —Si—O— | CH_3	−125
Polybutadiene	—CH_2—CH=CH—CH_2—	−85
Polyisobutylene (butyl rubber)	CH_3 | —C—CH_2— | CH_3	−70
Natural rubber	CH_3 | —CH_2—C=CH—CH_2—	−70
Polychloroprene (neoprene rubber)	Cl | —CH_2—C=CH—CH_2—	−50
Poly (vinyl chloride) (P.V.C.)	Cl | —CH_2—CH—	+80
Poly (methyl methacrylate) (perspex)	CH_3 | —CH_2—C— | C=O | O | CH_3	+100
Polystyrene	—CH_2—CH— | C HC⁄ \CH ‖ HC\ ⁄CH C H	+100

stress as well as on the temperature. This is the direct consequence of the reduced mobility of the chains, or high *internal* viscosity of the material. We are here concerned with a state in which the chains do not have the complete mobility of the ideal rubber, nor are they completely

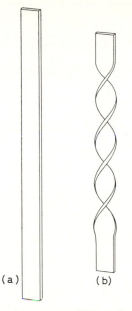

(a)   (b)

Fig. 5.7. (a) Original strip of perspex. This is heated above T_g and twisted (b). On re-cooling the twisted form becomes frozen in. When heated again the sample reverts to the original form (a).

set in position as in the glassy state; they can move to a limited extent, but their movement is accompanied by considerable drag or loss of energy. The deformation under a given stress is not a definite function of the stress, but depends also on its time of application; it is no longer truly *reversible*.

The easiest way of studying the mechanical behaviour in this region is by means of vibration experiments, which enable a definite frequency or time-scale to be assigned to the deformation. Suppose we take a sample of the material and subject it to a cycle of compression at a definite frequency, say 1 cycle per minute. Using fairly simple apparatus we may then measure both the amplitude of the stress (i.e. the difference between the maximum and minimum stress in the cycle), and the corresponding amplitude of the deformation. If we keep the stress amplitude constant, while the temperature is varied, the strain amplitude is found to vary in the manner shown in fig. 5.8. In the case of perspex, for example, at temperatures below about 110°C the amplitude is too small to be apparent on the scale of this diagram; above this temperature it rises very rapidly, but then begins to approach a limiting value. Below 100°C the material behaves like a glass, while above 150°C it behaves like a rubber; the intervening range represents the gradual transition from the one state to the other.

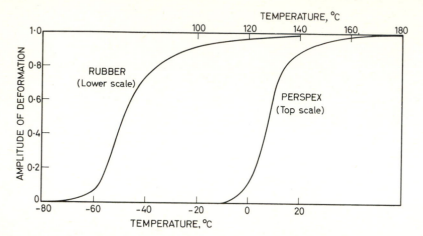

Fig. 5.8.   Temperature dependence on strain amplitude for rubber and
perspex in passing through transition region.

An alternative method of representing these properties is in terms of
the modulus of elasticity or Young's modulus E, at the particular frequency of vibration considered.   The modulus measured under these
conditions is called the *dynamic* modulus, to distinguish it from the so-
called *static* modulus obtained by applying a steady or static tension to
the specimen; its value is obtained from the ratio of the stress amplitude
to the strain amplitude.   A typical plot of dynamic modulus against
temperature for a glassy polymer (corresponding to the curve for perspex
in fig. 5.8) is shown in fig. 5.9.   In this figure the modulus is represented

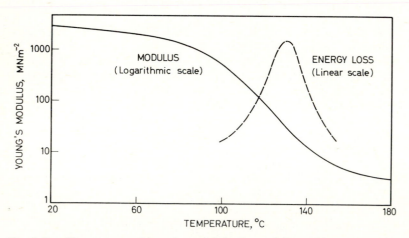

Fig. 5.9.   Temperature dependence of elastic modulus and energy loss for
glassy polymer.

74

on a logarithmic scale, in order to accommodate the large range of values covered. As the temperature is raised the modulus falls from a value of about 3×10^9 N m^{-2} (3 GN m^{-2}), which is typical of a glassy polymer, to about 3×10^6 N m^{-2} (3 MN m^{-2}), corresponding to a rubber.

In either the glassy or the rubbery state the material behaves elastically, i.e. the applied strain is completely (or almost completely) recovered on removal of the stress. In the intermediate region, however, the recovery is only partial. This means that the energy absorbed in the deformation part of the compression cycle is not recovered in the returning part; there is therefore a net loss or dissipation of energy in the cycle. The relative energy loss per cycle reaches a maximum at about the middle of the transition region, where the modulus is changing most rapidly (fig. 5.9). This loss of energy corresponds to the work done by the applied force in overcoming the *viscous* resistance of the material to the deformation; in molecular terms, it is the energy absorbed in dragging molecules past one another, and is of the same kind as the energy absorbed in the viscous flow of a liquid. The polymer thus behaves as if it possesses internal viscosity as well as elasticity. Materials which have this property are called *viscoelastic*; they possess both the rigidity or elasticity of a solid and the viscosity of a liquid, though this viscosity is exhibited not in the form of bulk flow, but in the form of a resistance to internal molecular rearrangements associated with the limited deformation of a network.

The variation in properties in the transition from the glassy to the rubbery state is very similar in all the *linear* polymers which do not crystallize, the only difference between one material and another being in the value of the temperature at which the transformation takes place. (In very highly cross-linked resins, having the structure represented in fig. 1.2 the full rubbery properties are, however, not well developed). In a good rubber the change in properties is completed at temperatures well below the range likely to be met with in actual use; in natural rubber, for example, it occurs between $-70°C$ and about $-20°C$.

An effective demonstration of the range of variation in mechanical properties in the transition from the glassy to the rubbery state may be obtained by winding a coil of rubber rod (or thick-walled tubing) of about 5 mm diameter on to a cylinder of diameter, say, 5 cm formed from a sheet of paper. The rubber is frozen to the glassy state by immersing the whole assembly in a wide Dewar vessel or Thermos flask containing liquid nitrogen. On removal, the paper cylinder (which is now brittle) is broken away, leaving a rubber coil (fig. 5.10 (*a*)) which retains its shape, and may be deformed like a spring by pulling between the hands. As the temperature rises, the spring deforms more easily, but the elastic recovery becomes more sluggish, until a stage is reached when the material becomes almost completely inelastic. In this region of temperature, the rod will retain any shape into which it is deformed (*b*). At still higher temperatures it becomes sufficiently flexible to hang

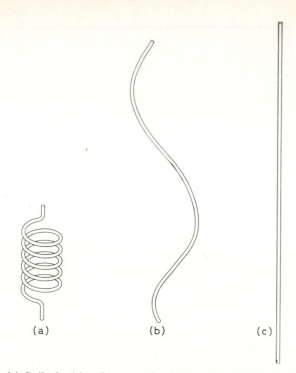

Fig. 5.10. (*a*) Coil of rubber frozen in liquid nitrogen (*a*) behaves like a spring. On warming it passes through a poorly elastic stage (*b*) before becoming straightened out (*c*).

straight under its own weight (*c*) but still stretches only slightly when pulled. The final stage corresponds to the recovery of the full rubbery elasticity.

This demonstration shows clearly the two distinct types of elastic response; the small-deformation, high-modulus type at low temperatures, and the large-deformation, low-modulus type at high temperatures. In between, we see the gradual variation of modulus, and the very large viscoelastic effects which characterize the transition region.

8. *Rebound resilience*

The rebound resilience or 'bounce' of a rubber ball changes very greatly as the temperature of the glass transition is approached. This can easily be understood from the preceding discussion of the energy loss in cyclic deformation. On impact, the ball suffers a local compression; this compression, and the recovery which follows, may be regarded as equivalent to half a complete cycle of oscillation, the time involved being of the order of a few thousandths of a second. The rebound resilience is defined as the ratio h/h_0, where h_0 is the height from which the ball is dropped, and h the height of the rebound. The difference, $1 - h/h_0$, is a measure of the relative energy loss in the half-cycle.

The variation of resilience with temperature (fig. 5.11) shows a minimum at the temperature at which the energy loss (fig. 5.9) is a maximum. For natural rubber this is at about $-35°C$. At this point the rubber will not bounce at all; it is completely 'dead'. The im-

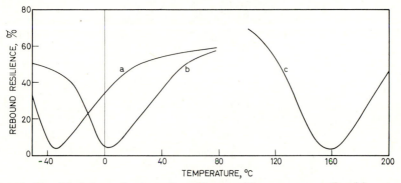

Fig. 5.11. Variation of rebound resilience with temperature for (a) natural rubber, (b) nitrile rubber and (c) perspex ((a) and (b) from Mullins, 1947, (c) adapted from Gordon, 1957)).

proved resilience at higher temperatures is associated with the greater mobility of the chain segments in the network. At temperatures below the position of the minimum the resilience also improves, but the deformation in this region is of the glassy, high-modulus type, not of the rubbery type. In this condition the sound emitted on impact is like that given out by a wooden ball when dropped on to a concrete floor.

The temperature of the minimum resilience is directly related to the glass-transition temperature. For example, in the case of the synthetic rubber known as nitrile rubber, which contains the CN-group attached to the chain to confer resistance to absorption of oils (see Chapter 10) the transition temperature is about $40°C$ higher than in natural rubber, and the minimum resilience occurs at $+5°C$ (fig. 5.11). This does not matter in most applications where this material is used as an oil-sealing material, but it would rule out its use in applications involving high rates of loading or vibrations, e.g. as an anti-vibration mounting.

A comparable curve for perspex shows exactly the same type of variation, but the minimum is now at $160°C$ (fig. 5.11).

9. The brittleness of glass

The subject of brittle fracture will be examined in detail in Chapter 9, when we come to consider the strength of materials, but before leaving the subject of the glassy state some reference must be made to this important topic.

A great deal of attention has been given to the problem of brittle fracture in inorganic glasses. Despite their apparent fragility, these glasses are not in fact particularly weak; the tensile strength of an ordinary lime-soda glass (as used for windows) is 8×10^7 N m^{-2}. This may be compared with rubber, whose tensile strength (referred to the unstrained dimensions) is about 2×10^7 N m^{-2}. Yet nobody would regard rubber as weaker than glass. The reason for the apparent fragility of glass lies not so much in its inherent lack of strength, as in its inability to deform beyond a certain very small limiting value of the strain; this *indeformability* is particularly disadvantageous under impact or shock loading conditions. The maximum extension which a glass will tolerate is of the order of only 0·1%; if the deformation exceeds this value the glass breaks. Strains of this order are easily produced by a relatively small amount of bending.

Most materials, particularly metals, are capable of larger *elastic* deformations, but what is more important, if they are subjected to strains exceeding the elastic limit they do not immediately break, but deform *plastically*, or flow. If a sheet of, say, mild steel is struck with a hammer it may be dented, but it does not fail completely; the denting corresponds to a local plastic deformation. This relieves the stress and so confines the damage to a small area. The important feature of a glass is that it is incapable of sustaining any kind of plastic deformation.

The reason for this difference is associated with the *crystalline* structure of the metal. In the crystal there are certain planes of atoms—the so-called *glide planes*, which allow sliding to take place under the action of comparatively low stresses. In a polycrystalline metal there will always be a proportion of crystallites oriented in such a direction that the prevailing stress can be relieved by such a sliding process, or plastic deformation. In a glass, on the other hand, there is no ordered arrangement and consequently no particular plane on which sliding can occur; if the maximum permissible strain is exceeded, the molecules separate and fracture occurs. It is typical of glassy materials that the fracture surfaces are generally *curved*; such surfaces are called *conchoidal*, like a cockleshell (Latin *concha*, shell), and reflect the form of the stress distribution at the moment of rupture. In a crystalline solid, the planes of fracture are related to the structure of the material; as a result the fracture surface usually presents a granular appearance.

The sensitivity of glass to rapidly created differences of temperature is another aspect of the same inability to sustain a large deformation. For a typical glass with a coefficient of linear expansion of $1·0 \times 10^{-5}$ per K, a temperature gradient of 100 K per cm through the thickness of the material will give rise to strains of the order of 0·1%, which are sufficient to cause failure. Hence, if put into an oven, ordinary glass will probably crack. Glasses for ovenware, such as pyrex, contain a high proportion of silica and boric oxide; their expansion coefficients are only about one-third of the above value. Pure silica glass has the extremely

78

low expansion coefficient of 0.5×10^{-6} per K, and is therefore practically immune to thermal shock; it may be exposed directly to the heat of a furnace.

A second very important characteristic of glasses is their susceptibility to crack propagation. As we shall see in Chapter 9, a sharp crack has the effect of concentrating or magnifying the applied stress. In the case of a metal the higher stress developed at the tip of a crack will lead to local plastic deformation; this automatically rounds off the tip of the crack and relieves the stress concentration. Provided the applied stress is not too high, this effectively prevents the spread of the damage by crack propagation. In a glass there is no mechanism for relieving the local stress, and a crack, once developed, continues to be propagated so long as the small amount of energy required for this purpose is available.

The polymeric glasses suffer from the same kind of brittleness as the inorganic glasses, but their brittleness is less extreme than in the case of inorganic glasses, for two reasons. Firstly, they are capable of somewhat larger elastic deformations, of the order of 1% rather than 0.1%. Secondly, they are not so *absolutely* rigid and incapable of local molecular movement as the inorganic glasses. Consequently the problem of brittle fracture in these materials, although a serious one, does not have quite the overwhelming importance that it has in the case of the inorganic glasses.

10. *Conclusion*

From the study of the nature of the glassy state and of the changes in physical properties which take place in the transition region we are able to obtain a better understanding of the close relationship, in terms of molecular structure, between materials of widely different character. Properties which at first sight appear irreconcilably different, such as rubber-like deformability and brittle fracture, are seen merely as extreme points in a continuous range of properties which may be exhibited in one and the same material as the temperature is varied. The realization of this continuity helps us to understand why a particular material has certain particular properties under any given conditions.

Although the glass-transition temperature has been treated as having a characteristic value for any given polymer, it must be emphasized that this is only true within limits. It is affected to some extent by the molecular weight, by the presence of chain branching, by cross-linking, and by other structural variations. It may also be greatly changed by the addition of low molecular-weight liquids. Thus polyvinyl chloride, which is normally a glass, with a transition temperature of 80°C, may be converted to a rubber by the incorporation of a non-volatile liquid or ' plasticizer ' such as di-methyl phthalate. The reason for this change can readily be deduced from the general picture of the structure of an amorphous polymer which has been presented. The presence of the plasticizer has the effect of reducing the number of contacts between polymer molecules, which is equivalent to reducing the strength of the intermolecular forces. This increases the mobility of the polymer chains, and hence reduces the temperature at which glass formation sets in.

This example is an illustration of the great versatility to be found not only among different types of polymer, but in any one polymer, according to the way it is processed or compounded, and the conditions under which it is used. This versatility confers both advantages and disadvantages. The great advantage is that the range of properties may be extended or adapted to meet particular requirements. The main disadvantage is that the problem of specifying and maintaining particular properties to satisfy a given practical requirement is considerably more complicated than is the case with the traditional materials of construction.

CHAPTER 6

crystallization phenomena in rubber

1. *Introduction*

IN preceding chapters references have already been made to the distinction between crystalline and amorphous polymers, and some indication has been given of what is implied in terms of molecular *arrangement*, or molecular structure, when crystallization takes place. In the present chapter we will take a closer look at the actual process of crystallization and try to understand some of the outstanding effects on the physical properties of a polymer which are brought about by this process.

Just as in the case of the glassy or amorphous polymers, it was found illuminating to start by examining the process of transformation to the glassy state in rubber, where its effects are rather more striking and more readily demonstrated, so also in the matter of crystallization a great deal may be discovered from an examination of the phenomena of crystallization as they may be observed in rubber. We will, therefore, begin our examination of crystallization and crystalline polymers with this material and will go on in later chapters to see how far the principles elucidated in the study of rubber may be applied to the interpretation of the structure and properties of the specifically crystalline polymers such as, for example, nylon and Terylene.

The advantage of using rubber as a kind of model substance for the study of crystallization is that in this material the processes of crystallization take place quite slowly, and within a convenient temperature range, whereas in most crystalline polymers they are extremely rapid and occur at high temperatures. The processes of crystallization in rubber can therefore be studied with greater ease, and subjected to more precise measurement than is possible with a typical crystalline polymer. Furthermore, by altering the temperature, or by introducing a strain into the material, it is possible to modify the process of crystallization in various ways, and hence to bring out effects which are of fundamental importance but less amenable to control in crystalline polymers generally.

2. *Crystallization by freezing*

It must first be made clear that the phenomena of crystallization are not observable in all types of rubbers. It is only those rubbers whose chains are made up of a regular repeating unit that are capable of crystallization. This applies to natural rubber and to gutta-percha, both of which are polyisoprenes of regular structure, and also to the synthetic rubber polychloroprene (Neoprene) (table, p. 72). It does not apply to most of the other synthetic rubbers, e.g. butadiene-styrene rubber and butyl rubber, which contain more than one type of monomer unit in the chain. Unless otherwise stated, the term rubber, as used in the present chapter, will be taken to mean *natural* rubber.

Natural rubber may be crystallized either by cooling below room temperature or freezing, in the normal unstrained state, or by stretching (at room temperature). We shall start by considering crystallization by freezing.

On being held at a low temperature, e.g. 0°C, for a period of several days, unvulcanized or raw rubber gradually loses its elasticity and becomes stiff and hard like a board. At the same time it loses its optical transparency and acquires a yellowish waxy appearance. The gradual hardening is quite different from the type of hardening which occurs on cooling quickly to a temperature below the glass-transition temperature. This is instantaneous, and in the glassy state the rubber is glass-hard and brittle. The changes in properties on crystallization are gradual and also less severe; though the increase in hardness is considerable, the material still retains a degree of flexibility. It can be bent between the fingers and is not brittle.

Unlike the transition to the glassy state, which corresponds to an instantaneous ' freezing-in ' of the amorphous structure of the rubber, crystallization involves a change in the actual structure or arrangement of the molecules. For crystallization to take place it is necessary for molecules—or more precisely, segments of molecules—to move relatively to one another, and to arrange themselves in a regular manner on a crystal lattice. This process of rearrangement or change of structure necessarily requires a certain amount of time for its accomplishment. This is why crystallization in rubber is so much slower than the transition to the glassy state, which represents only a suppression of the state of molecular *motion*, without any change in molecular structure.

In most respects the process of crystallization in a rubber is no different from the process of crystallization in an ordinary liquid on solidification. There are, however, important differences which are associated with the long-chain character of the polymer molecule. In a liquid the individual molecules can move quite freely, so that when the temperature is reduced to the freezing point, or below, the necessary rearrangement corresponding to the change from a disordered or irregular to an ordered or regular state of packing can take place very rapidly. In a polymer, on the other hand, the complex local entanglements between chains greatly reduce the mobility of the chain segments and render the process of rearrangement very much more difficult. The time-scale for crystallization in a polymer is therefore much longer than in an ordinary liquid of low molecular weight.

82

3. Rates of crystallization

From the above discussion it will be apparent that a detailed study of the rate of crystallization under varying conditions of temperature may provide important clues to the actual *mechanism* of crystallization, that is to say, to the manner in which the formation of a distinctive crystalline phase is actually brought about. Extensive studies on these lines were first made by Bekkedahl and Wood at the National Bureau of Standards in Washington, U.S.A. As with any other crystallization process, the crystallization of rubber involves a change in the density of the material; in the crystalline phase the molecules are more closely packed and the density is accordingly higher than in the original amorphous material. This change of density (or, strictly speaking, of volume) may be easily measured by means of a dilatometer of the type already referred to (fig. 5.5).

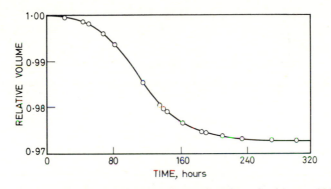

Fig. 6.1. Change in volume on crystallization of rubber at 0°C (Bekkedahl, 1934).

Figure 6.1 shows the changes in relative volume which take place during the crystallization of rubber at 0°C, determined in this way. The shape of this curve is quite characteristic. The rate of reduction of volume (or increase of density) is at first quite slow, but with increasing crystallization it accelerates and becomes relatively fast. In the later stages of crystallization, however, the rate again begins to slow down. Thereafter the process becomes slower and slower until it finally ceases.

The effect of temperature on the rate of crystallization is very striking. As the temperature is reduced the characteristic shape of the curve of crystallization remains unchanged, but the whole process becomes compressed into a shorter time-scale (fig. 6.2). Ultimately, however, a temperature is reached at which the rate passes through a maximum, so that with further reduction of temperature the initial trend is reversed. This is shown in fig. 6.3, in which the rate (measured by the reciprocal of the time taken to attain half the final change of volume) is represented

83

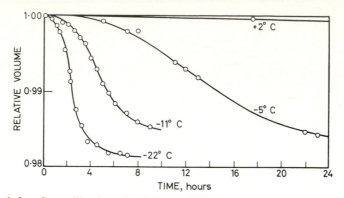

Fig. 6.2. Crystallization of rubber at different temperatures (Bekkedahl and Wood, 1941).

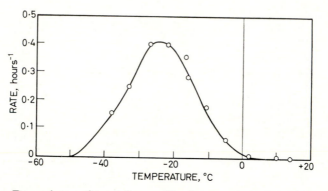

Fig. 6.3. Dependence of rate of crystallization on temperature (Wood, 1946). (The rate shown is the reciprocal of the time for crystallization to reach half its final value.)

as a function of the temperature of crystallization. The maximum occurs at a temperature of about $-25°C$. At this temperature the process is completed in about five hours.

4. Interpretation of rate curves

It is not difficult to account for the characteristic form of these crystallization curves. The initial delay in building up the rate of crystallization is associated with what is called *nucleation*. By this we mean that crystals can only grow and develop if there is already in existence a certain number of very small particles or nuclei which act as centres for the deposition of crystalline material. These centres may be thought of as exceedingly minute crystals containing only a small number of molecular segments which have come together in the course of their random thermal motions. These original nuclei are not very stable,

84

and in the course of time many of them will again disperse. The stability of the nucleus, however, increases with its size, and if a particular nucleus succeeds in surviving for a sufficient length of time for a number of other molecules to become attracted to its surface it may reach a size at which it is assured of a permanent life. Thereafter it will continue to grow by the deposition of further amorphous material on to its surface.

It may be assumed that the rate at which crystalline material is deposited on to the surface of a developing nucleus is directly proportional to its surface area. Hence as the nuclei becomes larger their rate of growth increases in proportion. This mechanism therefore accounts for the rapid increase in the rate of crystallization in the early stages of crystal growth.

Nucleation is not a peculiar feature of polymer crystallization but is a factor common to all processes involving the deposition of a separate phase from an initially uniform or homogeneous medium. It plays a dominant role, for example, in the formation of cloud or raindrops from atmospheric water vapour. In this case fine dust particles usually play the part of nuclei for condensation. In perfectly dust-free air very high degrees of ' supersaturation ' may be sustained. The reason for this is connected with the surface tension or surface energy of the liquid phase. The molecules in the surface layer of the liquid are in a state of higher energy than the molecules in its interior; consequently the formation of a fresh surface requires the expenditure of a definite amount of energy. It is therefore easier for molecules from the vapour phase to condense on to an already formed liquid surface (or for that matter on to any solid surface) than to form an entirely fresh surface. Exactly similar arguments apply to the separation of a solid phase from a liquid medium, and account for the well-known phenomena of the supercooling of a liquid below its equilibrium freezing point and the supersaturation of solutions.

In the case of rubber, the accelerating rate of growth of nuclei does not, however, continue indefinitely. As the growth proceeds the effect of the geometrical entanglements between chains attached to neighbouring crystallites becomes a more serious factor in the situation and the freedom of motion of individual chain segments becomes progressively more and more restricted. This will be apparent from fig. 6.4 (a), in which the crystallites are represented by the regions of parallel alignment of chains. As a result of this geometrical restriction or diminished chain mobility, it becomes increasingly difficult for chain segments to get into a suitable position for attachment to one or other of the already formed crystallites. A stage is eventually reached where further crystallization can only proceed at a rapidly diminishing rate until ultimately it effectively ceases.

Because of these geometrical entanglements the process of crystallization in a polymer never proceeds to completion: it always stops well short of 100% crystallization. The precise amount of crystallization varies with the type of polymer, and also to a limited extent according to the

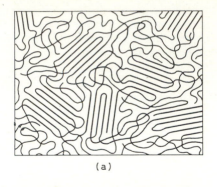

(a)

(b)

Fig. 6.4. Structure of rubber crystallized (*a*) in unstrained, and (*b*) in strained state. The ordered domains represent individual crystallites.

conditions of crystallization. In the case of rubber the proportion of crystalline material present in the final state is usually not more than about 30%.

The effect of temperature on the rate of crystallization can also be explained in terms of the two factors, nucleation and chain mobility. Reduction of temperature increases the probability of formation of a stable nucleus, the reduction of the thermal energy of the chain segments making it less likely that a nucleus, once formed, will disperse. The number of nuclei and hence the overall rate of crystallization thus increases as the temperature is lowered. At sufficiently low temperatures, however, the amount of energy available to the chain segments becomes so greatly diminished that it seriously interferes with their mobility. Under these conditions the dominant factor in the situation is the chain mobility, which progressively diminishes with further reduction of temperature. It is noteworthy that at temperatures below $-50°C$ the rate of crystallization becomes too small to be observable. This is some $20\,\text{K}$ from the glass-transition temperature ($-70°C$), where, of course, all segmental motion is completely frozen out and no further change of structure is possible.

5. *Melting phenomena*

By ' melting ' in the case of a rubber we mean the transition with rise of temperature from the relatively hard crystalline state to the soft rubbery or amorphous state.   As with crystallization, the phenomena of melting in a polymer are more complex and more varied than in a low molecular-weight material.   Figure 6.5 shows the changes in volume

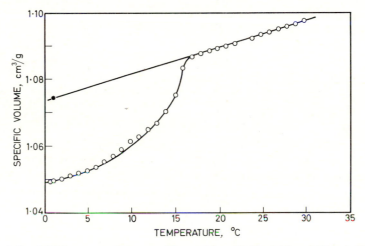

Fig. 6.5.   Melting curve for crystalline rubber (Bekkedahl and Wood, 1941).

which take place when crystalline rubber is slowly heated.   There is at first a steady linear increase in volume corresponding to the ordinary thermal expansion of the material.   At a certain temperature, corresponding to the beginning of melting, the curve begins to rise more steeply. This rise continues over a range of some 10°C (for rubber crystallized at −2°C) until melting is complete.   After this the normal expansion of the amorphous rubber is observed.   If, after melting, the temperature is again reduced, the linear reduction of volume for the amorphous rubber is continued down to lower temperatures, since appreciable recrystallization is not observed on the time-scale of this experiment.

The first noticeable difference, compared with a low molecular-weight material, revealed by experiments of this kind, is that the melting temperature of crystalline rubber is not sharp, but covers a considerable range.   The second, and more surprising difference, illustrated in fig. 6.6, is that both the temperature at which melting starts and the range of temperature over which it extends depend on the temperature at which the preceding crystallization has been carried out.   Melting does not start until the temperature is raised to a value from four to six degrees in excess of the temperature of crystallization.

Bearing in mind the fact that in ordinary chemical compounds the

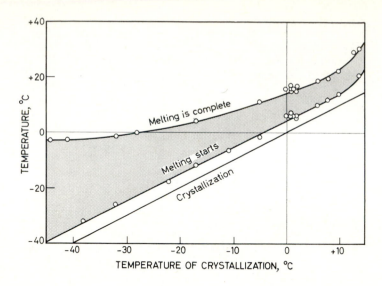

Fig. 6.6. Dependence of range of melting on temperature of crystallization (Wood, 1946).

melting point is one of the most accurately specifiable of its properties, these variations in melting properties in rubber are indeed remarkable. They must be interpreted as a failure of the system to attain a genuine state of equilibrium between the crystalline and amorphous phases. In an ordinary liquid in contact with the solid phase at the melting point all the molecules in the liquid phase can interchange freely and move to any point on the surface of any crystal; all regions of the liquid are equivalent. Such freedom does not exist in a polymer, where the chain segments are physically connected to the crystallites and where the same molecule may extend from one crystallite to another via the intervening amorphous region (fig. 6.4 (a)). In such a system the individual crystallite is not, and cannot be, in equilibrium with the amorphous phase *as a whole*; it is only in equilibrium with the molecules, or molecular segments, which happen to be in its immediate vicinity. The types of entanglement and local chain conformations will vary from one region to another in the structure. Such a system is not homogeneous in properties and does not behave in the regular manner of a normal liquid in equilibrium with a solid phase.

This failure to attain a true equilibrium gives rise to some rather unusual effects. Inspection of fig. 6.6 shows that for rubber crystallized, for example, at $-40°C$, melting is complete at about $-4°C$. But if after melting of the crystals has been carried out the amorphous rubber is maintained at this same temperature ($-4°C$) for a longer period, a second crystallization process will commence. Under other conditions

88

it is possible to have both melting and recrystallization processes taking place simultaneously in the same piece of rubber!

Another effect of interest is that the melting temperature of crystalline rubber increases progressively with increasing time of storage. After many years storage in a cool cellar it may be as high as 39°C. If one half of a sheet of such crystalline rubber is converted to the amorphous form by heating, this half will not recrystallize at normal living-room temperatures (say ~ 15°C), so that both amorphous and crystalline portions will remain in these respective states for an indefinite time*. Figure 6.7 illustrates the effect of applying a tensile stress to such a specimen; the highly elastic extension is confined to the amorphous half.

6. *Crystallization by stretching*

It has been known for a long time that when a strip of vulcanized rubber is stretched it becomes crystalline, and that this crystallization

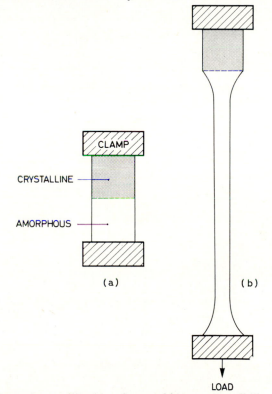

Fig. 6.7. Effect of crystallization of raw rubber on extensibility. (*a*) Original sample, top half crystalline, lower half amorphous. (*b*) Extension is confined to amorphous portion.

* The author has in his possession a sample of this kind which has preserved its double state unchanged for more than thirty years.

immediately disappears on removal of the stretching force. Stretched crystalline rubber yields a characteristic type of X-ray diffraction pattern (Plate 1 (*b*)) known as a 'fibre diagram'; this is the type of pattern produced when the axes of the crystallites are all lined up parallel to a single direction, corresponding to the axis of the fibre.

At first sight it seems very difficult to understand how the mere process of extension can actually produce crystallization in a rubber, and more particularly why such crystallization should be reversible. As it turns out, vulcanized rubber itself is not the most suitable system for the investigation of this phenomenon of crystallization by stretching. In unvulcanized rubber the same phenomenon is manifested in a rather more striking manner, and may be examined under a greater variety of conditions.

Figure 6.8 shows the changes in density which take place when raw rubber is held at various fixed extensions at a temperature of 0°C. In

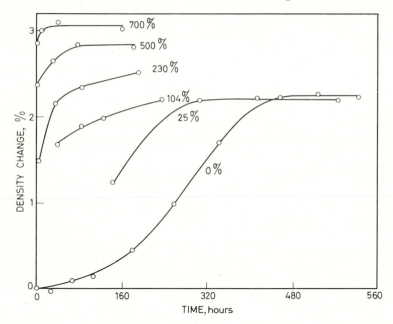

Fig. 6.8. Changes in density in rubber held at the extensions indicated at 0°C.

these experiments strips of rubber were stretched to the required extension and clamped to a board. After various times samples were cut out for the measurement of density (by hydrostatic weighing), also at 0°C. The curve for the original unstrained rubber (zero extension) has an equivalent form to the corresponding curve of relative volume (fig. 6.1), an increase of density being equivalent to a reduction of volume. Moderate extensions are seen to have a similar effect to a reduction of

90

temperature; the rate of crystallization is increased, but the general form of the crystallization curve (in so far as it can be observed) remains unchanged. At higher extensions the rate becomes so high that only the final stages of the process are observable. Under these conditions crystallization *appears* to take place simultaneously with the extension.

An alternative method of following the course of crystallization makes use of the phenomenon of *double refraction*. This phenomenon is commonly observed in crystals; it denotes the property of splitting of a ray of light incident on the surface into two refracted rays which travel through the medium with different velocities. These two rays are *polarized* in planes at right angles to each other. This implies that the refractive index of the crystal depends upon the plane of polarization of the light, that is to say, on the direction of the electric vector in the electromagnetic wave.

The crystallites present in stretched rubber are themselves doubly refracting, and being lined up parallel to the axis of extension, they confer corresponding doubly refracting properties on the rubber. (There may be an additional small amount of double refraction due to the mere alignment of the molecules in the amorphous regions of the structure, but this need not concern us here). The resultant double refraction may readily be observed in polarized light, and its amount may be expressed in terms of the difference of refractive index for light polarized respectively parallel and perpendicular to the direction of extension.

Experimental results obtained by this method are shown in fig. 6.9, in which are plotted the changes in double refraction in the course of time, for various fixed extensions. The method has the advantage that measurements may be made more rapidly than can density measurements, and the sample does not have to be removed from the clamps for the purpose of making the measurement. It is therefore possible to follow the process of crystallization in its earlier stages. On the other hand, it is not possible to make observations on the unstrained rubber, because in this state the axes of the crystallites have no preferred orientation, and hence, although the individual crystallites are doubly refracting, the whole assembly is effectively uniform in optical properties.

The results shown in fig. 6.9 confirm the density observations. The conclusion to be drawn is that the effect of extension is to increase the rate of crystallization, and also to some extent the amount of crystallization, without in any way changing the general nature of the crystallization curves. The inference is that the process of crystallization in the stretched rubber does not differ in principle from the corresponding process in the unstrained material. The effect of the extension is to produce a degree of molecular alignment (cf. fig. 4.1). This will increase the probability of formation of a nucleus, the nuclei now being formed preferentially in the direction of the extension, and at the same time produce a more favourable disposition of chain segments for further crystallization on these oriented nuclei. Both of these effects will there-

91

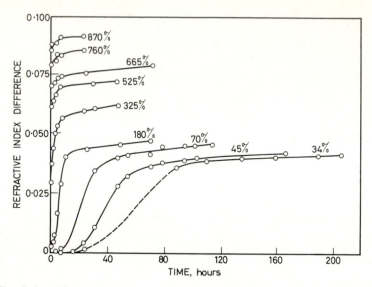

Fig. 6.9. Changes in double refraction in rubber held at the extensions indicated at 0°C.

fore tend to produce an increase in rate of crystallization with increasing extension. The resulting structure may be envisaged as having the form represented in fig. 6.4 (*b*), the degree of alignment of the crystallites becoming more perfect with increasing extension.

These experiments enable us to bridge the gap between the very slow crystallization of unstrained rubber and the apparently instantaneous crystallization of the highly stretched material, which at first sight appears to be a different phenomenon. These two processes are now seen to be the two extremes of a continuous gradation. Though crystallization may be very fast it is never instantaneous. The direct effect of the extension is to produce an orientation of the molecules; the increased rate of nucleation and crystallization is a consequence of this orientation.

7. *Mechanical properties of stretched crystalline rubber*

In vulcanized rubber the crystallization produced by extension is not permanent but disappears as soon as the stretching force is removed, when the material reverts to its original unstrained condition. Raw rubber, however, when crystallized at a high extension, remains in the extended state and does not return to its original state unless subsequently heated.

This effect may be readily demonstrated. A strip of raw rubber is stretched between the hands to a high extension and held in the stretched state for a minute or so. (It may be desirable, in a warm room, to cool it in a stream of tap water, or by contact with a beaker of ice-cold water).

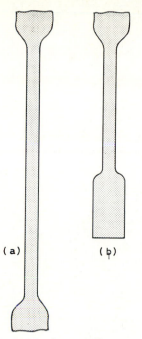

Fig. 6.10. Properties of stretched crystalline raw rubber. (*a*) Extended crystalline state. (*b*) Retraction of lower portion to original state on 'melting' of crystallites.

On removal of the stretching force it remains in the fully stretched state (fig. 6.10 (*a*)). On dipping into hot water, however, the crystals melt, whereupon it reverts to the unstrained state (fig. 6.10 (*b*)).

The difference in recovery behaviour between vulcanized and unvulcanized rubber after crystallization may be explained by the presence of a permanent cross-linked network structure in the vulcanized material. This network structure leads to a strong elastic restoring force which is sufficient to break down the crystalline structure formed on stretching. In unvulcanized rubber molecular slippage can take place on extension, and the elastic restoring force is not sufficient to break down the crystalline structure. It is only when the temperature is raised sufficiently to melt the crystallites (i.e. to about 30°C) that the rubber returns elastically to its original state.

In the highly extended crystalline state raw rubber has properties resembling those of a fibre. Its tensile strength in the direction of the extension is much greater than in the transverse direction. If a small longitudinal cut is made in a strip of highly extended crystalline rubber, it may be readily torn down the whole length of the specimen. If the same material is cooled in liquid nitrogen and struck with a hammer, a fibrous type of fracture, reminiscent of the fracture of wood, is produced

93

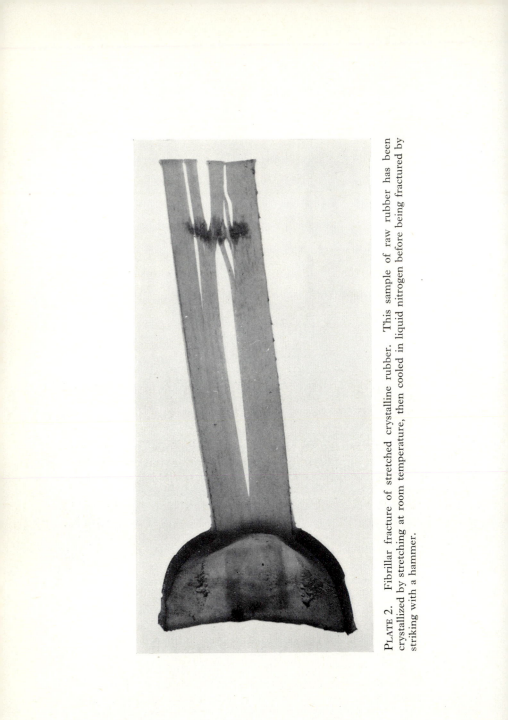

PLATE 2. Fibrillar fracture of stretched crystalline rubber. This sample of raw rubber has been crystallized by stretching at room temperature, then cooled in liquid nitrogen before being fractured by striking with a hammer.

(Plate 2). This difference in properties in different directions, or mechanical *anisotropy*, as it is called, is a direct consequence of the high degree of molecular orientation in the stretched crystalline state.

8. Conclusion

This survey of the phenomena of crystallization in rubber brings out the remarkable diversity of properties exhibited under different conditions of temperature and varying states of strain. Many of these properties are not at all well known, since they do not come within the range of normal industrial experience. The elucidation of these various phenomena, historically, was not achieved all at once. It was as early as 1925 that Hock first drew attention to the fibrous form of fracture exhibited by highly stretched raw rubber (then known as ' racked ' rubber) and attributed it to crystallization. His insight was confirmed a year later by the X-ray observations of Katz, which provided the first direct evidence for the existence of crystallization in rubber. It was only after many years, however, that the interpretation of these phenomena in terms of the molecular structure of the material gradually became apparent.

The importance of the phenomena considered in this chapter lies not so much in the bearing which they have on the properties of rubber as such, as in the insight they yield into the processes of crystallization in polymers generally, and into the principles underlying these processes. The effects of molecular orientation, in particular, and the relation between mechanical properties and molecular orientation, can be more readily isolated and studied in rubber than in many other polymers. These effects are of great importance in the study of fibres and the processes of fibre formation, and their elucidation has proved to be a valuable stepping-stone in the development of ideas on the structure and properties of these more complex materials.

CHAPTER 7
crystalline polymers

1. *Mechanical properties of crystalline polymers*

CRYSTALLINE polymers—polyethylene, polypropylene, nylon, etc.—have properties which are intermediate between those of the rubbers on the one hand and the glassy polymers on the other. They are considerably harder than rubbers, yet flexible and tough, not brittle like a glass. In the form of fibres they are among the strongest materials known. This combination of properties gives them advantages over other types of polymers and non-polymeric materials in a rapidly growing range of practical applications.

The properties of crystalline polymers are directly related to their peculiar type of structure. Their general structure is similar to that of crystalline rubber, which is illustrated in fig. 6.4. The individual crystallites, which are comparatively hard and undeformable, are, so to speak, embedded in an amorphous or rubber-like medium. The rubbery constituent confers some degree of rubber-like deformability, rendering the system flexible rather than brittle, but the crystalline component greatly modifies the rubbery properties and produces a type of material which is considerably harder and stiffer than a rubber.

Crystalline polymers differ from ordinary crystalline materials in several important respects. In most ordinary crystalline materials, e.g. metals, any individual crystal is in contact with other crystals at all points on its surface boundary; the system consists of a solid mass of crystals with no appreciable amount of disordered material in the inter-crystalline boundaries. The properties of the mass are therefore substantially the properties of the crystals themselves. In a crystalline polymer not only is there a considerable fraction of material in the amorphous or disordered state, but this material is formed of the *same* molecules as those which are present in the crystalline component. The system is thus not *just* a mixture of crystalline and amorphous material, but a combined structure made up of crystalline and amorphous components mechanically connected together in a very special way.

The hardness of a material may be represented in terms of the value of Young's modulus, which is the ratio of the applied stress (i.e. force per unit area) to the corresponding deformation or strain. Typical values of this quantity for various materials are shown in fig. 7.1. At one extreme we have the inorganic crystalline materials, diamond, quartz, steel, etc., with moduli in the range 10^{11} to 10^{12} N m^{-2}. These are

96

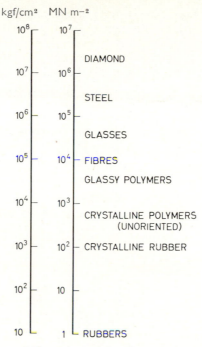

Fig. 7.1.   Values of Young's modulus for various types of materials
(logarithmic scale).

closely followed by the inorganic glasses, whose moduli lie between 10^{10} and 10^{11} N m^{-2}. For the glassy polymers the moduli are somewhat lower, 10^9 to 10^{10} N m^{-2}. The common crystalline polymers (in the unoriented state) are down by a further factor of 10, and fall in the range 10^8 to 10^9 N m^{-2}, with crystalline rubber at the lower end of this range. In the oriented form, as fibres, however, they have moduli about equal to, or a little higher than those of the glassy polymers. At the extreme low end of the scale come the rubbers, which are in a class by themselves, with moduli of about 10^6 N m^{-2} or less.

This enormous range of properties, represented by a factor of more than a million from the lower extreme to the top of the scale, reflects the different types of structure to be found in these various materials and the different kinds of forces which hold the molecules together within the structure. The hardest materials, such as diamond, are those in which every atom is connected to each of the surrounding atoms in the structure by strong primary chemical bonds of the same kind as those responsible for the formation of chemical compounds. These materials are essentially 3-dimensionally bonded structures. In the glassy polymers the forces which maintain the structure are the weaker secondary forces *between* molecules; this is why such materials have lower values of elastic

97

modulus. In rubbers, as we have already seen, the mechanism of elasticity is of an entirely different type, giving values of modulus of a lower order of magnitude. In the crystalline polymers the molecules do not have the same freedom of movement as in a rubber, but they are not so strongly held as in the glassy state; the moduli of crystalline polymers are therefore intermediate between the values which are characteristic of these two classes of materials.

Some appreciation of the significance of the vast range of numerical values of moduli covered by the various classes of materials given in fig. 7.1 may be gained from the pictorial representation in fig. 7.2.

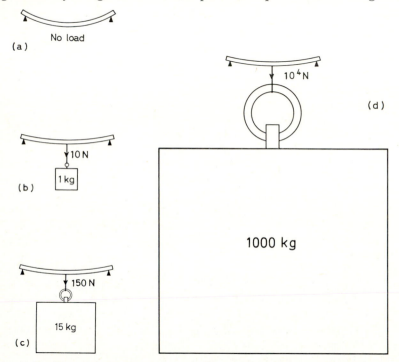

Fig. 7.2. Pictorial representation of differences in elastic modulus (stiffness) between different classes of materials. (*a*) Rubber. (*b*) Polythene. (*c*) Perspex. (*d*) Steel.

A bar of polythene of length 20 cm, cross-section 1 × 1 cm and modulus, say, 2×10^8 N m^{-2} would require a load of mass 1 kg (i.e. a force of 10 N) to produce a deflection of 1 cm at its mid-point. A similar bar of perspex (modulus 3×10^9 N m^{-2}) would require a load of mass 15 kg to produce the same deflection, while for steel (modulus 2×10^{11} Nm^{-2}) the corresponding load would be 1000 kg (10^4 N). A bar of rubber of the same dimensions, however, would deflect by 2·5 cm under its own weight!

98

2. Structure of the polymer crystal

For a better understanding of the process of crystallization in a polymer, and of the relation between the crystalline and amorphous components in the resulting structure, it is desirable to know something about the structure of the individual crystal itself, that is to say, about the precise way in which the molecules are arranged within the crystal. As with any other crystal, this arrangement may be described in terms of a small basic unit, the *unit cell* as it is called. This unit may be regarded as a kind of building brick which is repeated in all directions throughout the structure to produce a 3-dimensional lattice, in the same way as the pattern of a wallpaper is produced by the repetition in two dimensions of a small unit which constitutes the basic design. The determination of the dimensions of the unit cell is obtained by an analysis of the positions of the crystalline reflections in the X-ray diffraction pattern of the polymer, using the Bragg formula (p. 63). In order to obtain a suitable diffraction pattern for this purpose it is necessary to produce as much regularity as possible in the geometrical arrangement of the crystallites; this is achieved by a stretching or 'cold drawing' process. In the drawn state the axes of the crystallites are lined up in a direction parallel (or nearly parallel) to the axis of extension, and the diffraction pattern obtained is of the type known as a 'fibre diagram' (Plate 1 (*b*) and (*d*)). This type of pattern yields much more information than the 'powder' type of diagram obtained from the polymer in the undrawn or unoriented state (Plate 1 (*c*)).

The problem of deriving the dimensions of the unit cell is not too difficult, but the further analysis leading to the determination of the complete structure, that is to say of the exact geometrical disposition of the molecules within the unit cell, is far more complicated, and involves the consideration of the relative intensities, as well as the angular positions, of the various 'spots' in the diffraction pattern. Despite the difficulties, however, quite a number of polymer structures have been analysed, notably by Bunn, who was the first to work out the simplest structure of this type, the structure of the polyethylene crystal, in 1939. This structure is represented in fig. 7.3. The unit cell is a rectangular block having dimensions 0·253 nm (2·53 ångström) in the chain direction, and $0·740 \times 0·493$ nm in the transverse plane. The axial dimension of the unit cell corresponds to the distance taken up by *two* carbon atoms in the chain backbone structure; this is the *geometrical*, as distinct from the *chemical* repeating unit in the zig-zag chain structure, and corresponds to the distance $C_1 C_3$ or $C_2 C_4$ in the diagram below.

$$C_1 \diagup \overset{C_2}{} \diagdown C_3 \diagup \overset{C_4}{} \diagdown C_5 \qquad (7.1)$$

If we look down through the unit cell along the direction of the chain axes, we obtain the projection of the chains in the unit cell. This is

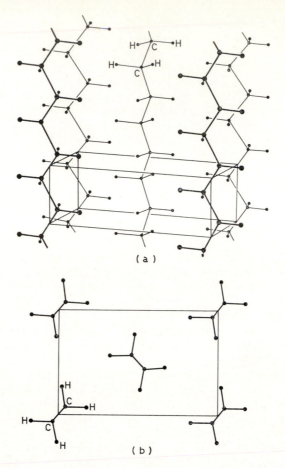

Fig. 7.3. Structure of polythene crystal, showing position of chains in unit cell. (a) Side view. (b) Projection along direction of chain axes (Bunn, 1939).

shown in the lower portion of fig. 7.3. In this diagram we see the C–C bond, which appears shortened in this projection, together with the two C–H bonds attached to each carbon atom. These C–H bonds lie in a plane at right angles to the chain axis, i.e. in the plane of the diagram.

Figure 7.3 represents only the *skeleton* structure of the molecules. This is the most useful form of representation for displaying the geometrical disposition of the molecules in the unit cell, but it gives an erroneous impression of the space occupied by the molecules, and of the way they are packed together. A more realistic impression of the packing of the molecules in the unit cell is conveyed by the lateral view in fig. 7.4 (top), in which the carbon atoms only are represented, and the

100

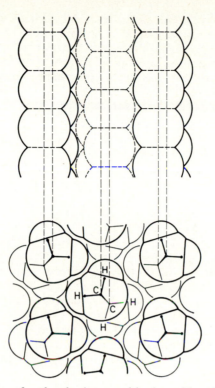

Fig. 7.4. Packing of molecules in crystal lattice. Top, lateral projection (H atoms not shown). Bottom, axial projection.

longitudinal projection in fig. 7.4 (bottom), which includes both the carbon and the hydrogen atoms. The beautiful way in which the parallel chains fit together is particularly well shown in this figure.

The general features of the structure brought out in the accompanying figures are quite typical. Another example, nylon, which is discussed in the following chapter, shows very similar features. The details of the structure, unit cell dimensions, etc., vary from one polymer to another, but in all cases the chains run parallel to one another through the unit cell and are precisely arranged with respect to one another in the transverse plane.

It is important to note that the analysis of the crystal structure is not concerned directly with the *polymeric* nature of the molecule as such. The crystal must contain a sufficient number of repeating units to yield a satisfactory diffraction pattern, but apart from this the length of the chain does not come into the problem. Nor does a knowledge of the structure of the crystal enable us to say whether the chains which lie side by side in the unit cell are part of one and the same molecule, or whether they

101

belong to entirely different molecules. All such questions have to be decided on other grounds. In fact, the dimensions of the unit cell, and the arrangement of the molecules within the unit cell for one of the low molecular-weight paraffins, containing, say, thirty to forty carbon atoms, are almost exactly the same as for the polythene crystal.

3. *Factors favouring crystallization*

The advantage to be gained from a knowledge of the crystal structure is that it helps us to understand the factors which are important in determining whether a polymer having a particular chemical structure will crystallize or whether it will be of the glassy or of the rubbery type. Some aspects of this question have already been touched on in earlier chapters, where it has been pointed out that the basic requirement for crystallization is regularity of chain structure. It is clear from what we have seen of the crystal structure that any lack of regularity in the individual chains will destroy the possibility of fitting them into a succession of identical unit cells. An extreme example of such irregularity is provided by copolymers, such as the butadiene-styrene rubbers, formed by the random addition of two different monomer units, but many other kinds of irregularity are also to be found. For example, if the repeating unit in the chain is itself unsymmetrical, having a ' head ' and a ' tail ', it is essential for crystallization that all the units in the chain should have their heads pointing in the same direction. This may be illustrated by natural rubber, in which the successive isoprene units are all joined ' head-to-tail ', as below:

$$-CH_2-CH=C-CH_2-CH_2-CH=C-CH_2- \qquad (7.2)$$
$$\qquad\qquad\quad | \qquad\qquad\qquad\qquad\quad |$$
$$\qquad\qquad\quad CH_3 \qquad\qquad\qquad\qquad CH_3$$

the methyl groups being either all at the right-hand end or all at the left-hand end of the repeating unit. In the synthetic polyisoprenes as normally produced the pendent methyl groups lie indiscriminately to the left or to the right. These materials cannot be crystallized either by cooling or by stretching, and consequently have significantly different mechanical properties from natural rubber. It is only recently that catalysts for the polymerization reaction have been discovered which yield perfectly regular head-to-tail addition; the ' synthetic natural rubber ' made in this way crystallizes and has all the properties of natural rubber itself.

Another type of irregularity which tends to reduce, and in an extreme form may completely suppress crystallization is chain branching. Each point of branching of the polymer molecule constitutes an irregularity which cannot be fitted into the crystal lattice without distortion of the structure. In the case of polythene, for example, we have for simplicity considered the molecule to consist of a single chain without branches. This is not strictly true of ordinary polythene, which is known to contain an appreciable number of branches or side chains, most of which are quite short. Again, by means of special catalysts, it is possible to produce a different type of polythene which consists entirely of completely unbranched chains. This form of polythene has a higher degree of crystallinity and consequently a higher density than ordinary 'low-density' polythene (see §6 below).

In relation to crystallization cross-linking has a very similar effect to chain branching, each point of cross-linkage being geometrically equivalent to a multiple branch point. It is for this reason that vulcanized rubber crystallizes very much more slowly than raw rubber at low temperatures, though with the normal small amount of cross-linking the reduction of the final degree of crystallinity is only slight.

4. Isotactic polymers

A more subtle form of irregularity may arise in certain types of polymers as a result of different configurations of the atoms attached to neighbouring carbon atoms in the chain backbone. This may be illustrated by the case of polypropylene, for which the repeating unit is represented by the formula

$$(7.3)$$

Considered as a structure in three dimensions, this unit may assume two different forms, as shown in figs. 7.5 (a) and 7.5 (b). In these figures

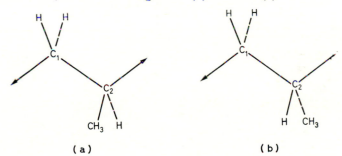

(a) (b)

Fig. 7.5. Alternative configurations of repeating unit of polypropylene chain. The dotted lines represent bonds *behind* the plane of the diagram.

the bond C_1–C_2 forming the chain backbone is considered to lie in the plane of the paper. Attached to C_1 are two H-atoms, of which one projects forwards, above the plane of the paper, while the other, for which the corresponding bond is represented by the dotted line, lies below the plane of the paper; these two H-atoms are of course equivalent. But at

103

C$_2$ the attached CH$_3$ group may lie either in front, as in (*a*), or behind, as in (*b*). The form (*b*) cannot be converted into (*a*) by any possible internal rotation about bonds; these two forms are distinct chemical structures. The one type of structure is the mirror image of the other, the two forms bearing the same relation to each other as the right hand bears to the left. Structures related to each other in this way are very common in organic chemistry and are known as spatial or *stereo*-isomers.

The normal process of polymerization produces a random sequence of (*a*) and (*b*) forms, as indicated in fig. 7.6 (*a*). A polymer of this kind is called *atactic*, meaning *without arrangement* (from the Greek *a*, without, taktikos, arranged)*. Because of their structural irregularity atactic polymers are incapable of crystallization. Atactic polypropylene is an amorphous rubbery polymer.

Comparatively recently types of catalyst have been discovered which enable polymers to be produced with the *same* configuration of successive groups along the whole length of the chain. Such polymers are known as *isotactic* (i.e. of the *same arrangement*). Isotactic polypropylene has the structure shown in fig. 7.6 (*b*). It is a highly crystalline polymer, similar in properties to polythene, and in the drawn state yields a fibre with excellent properties, which is being produced commercially in large quantities.

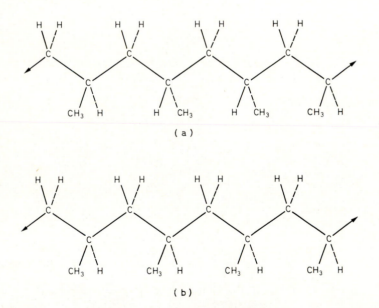

Fig. 7.6. Alternative structures of polypropylene chain. (*a*) Atactic. (*b*) Isotactic.

* *Tactics* is concerned with the *arrangement* of troops on the ground.

104

The same considerations apply to most of the glass-forming polymers. In polystyrene, for example, the same basic type of structure occurs, the repeating unit being represented by the formula

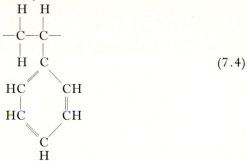

$$(7.4)$$

in which the benzene ring $-C_6H_5$ replaces the methyl group $-CH_3$ in polypropylene. As normally prepared, polystyrene has the irregular, atactic, type of structure and is therefore incapable of crystallization. An isotactic form of polystyrene has recently been produced, and this, as would be expected, can be crystallized. Unlike the isotactic form of polypropylene, however, this isotactic polystyrene does not appear to have properties of immediate value for commercial exploitation.

The other glassy polymer of major commercial interest, polymethyl methacrylate, having the repeating unit

$$(7.5)$$

also has the atactic form. The same applies to polyvinyl chloride

$$(7.6)$$

which may be used either as a glass, or, when mixed with a plasticizer, as a rubber. In either form it is non-crystalline.

We see, therefore, that this question of tacticity is one of the most important considerations in determining whether or not a polymer is capable of existing in the crystalline state. The atactic polymers are necessarily amorphous, and may be either rubber-like or glassy, according to the strength of the intermolecular forces.

5. Crystallization and melting phenomena

In the last chapter we examined in some detail the phenomena of crystallization and melting as they are observed in rubber. Observations of the same kind on typical crystalline polymers are not so readily made, chiefly because of the much greater rates of crystallization and the higher temperatures involved. The information which is available, however, indicates that the basic phenomena are essentially similar.

In fig. 7.7 are shown some observations on the rate of crystallization of polythene at various temperatures, obtained by measurement of volume changes. These curves are of the same form as the corresponding curves

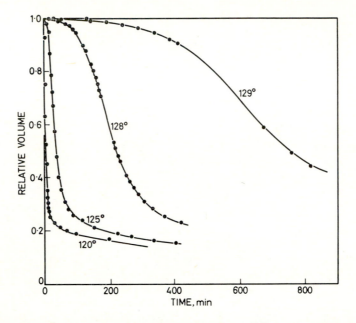

Fig. 7.7. Rate of crystallization of polythene at different temperatures, as indicated by volume changes (Mandelhern, 1964). The quantity plotted is the ratio of the actual volume change to the ultimate volume change after an indefinitely long time.

for rubber (fig. 6.2), but the time scale is very much shorter. As the temperature is reduced below the melting point the rate of crystallization rapidly increases, and at temperatures below about 115°C it becomes so rapid that observations are no longer possible. The form of these curves, and the effect of temperature, suggest that, as in the case of rubber, crystallization takes place about nuclei, and that the rate of formation of nuclei is the most important factor in the process. Again, by analogy with rubber, it is presumed that as the temperature is further reduced the rate will not continue to increase indefinitely, but will pass through a maximum and subsequently fall. This, however, is only a *presumption*, since it is not possible to make observations at these lower temperatures.

The very high rate of crystallization in polythene has important

106

practical consequences. It means that this material cannot be obtained in the amorphous state at room temperature. Even if 'shock-cooled' by plunging into water from the molten state, it is still crystalline, though the crystallites may be smaller and the amount of crystallization somewhat less. The time required for crystallization is shorter than the time taken in cooling the specimen.

The phenomena of 'melting' in crystalline polymers are likewise closely similar to the corresponding phenomena in rubber. Figure 7.8 shows a typical melting curve for 'high-density' polythene. As in the

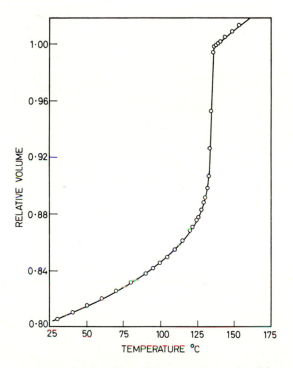

Fig. 7.8. 'Melting' curve for polythene as determined from volume changes (Mandelhern, 1964).

case of rubber, there is no well-defined crystal melting point, the process covering a range of temperature of some 10 K. Unlike rubber, however, the melting temperature for crystalline polythene does not vary to any great extent according to the temperature of crystallization, the maximum variations being limited to a range of two or three degrees at most.

The readiness of polythene to crystallize is probably related to the simplicity of its chain structure and the ease of packing of the chains in the unit cell. The rounded form of the cross-section of the chain

107

(fig. 7.4) and the absence of any projecting side-groups will facilitate both the local movement of chain segments and the formation of a regular packing. These properties, together with the absence of any strong intermolecular forces, are exhibited in a more pronounced manner in polythene than in most other polymers, and give us reason to regard polythene as a rather extreme type of polymer in relation to crystallization.

Terylene shows some similarities, but also some important differences from polythene in respect of its crystallization behaviour. The repeating unit in the Terylene chain (fig. 8.3, p. 120) is longer, more bulky and altogether more complex than the repeating unit in polythene. All these factors will make it more difficult for the chains to move into their correct positions in the unit cell, and would therefore be expected to reduce the rate of crystallization. These expectations are borne out in practice; the rate of crystallization in Terylene is sufficiently slow to enable it to be cooled to room temperature without crystallizing. In this state, however, it is glassy rather than rubbery, the glass-transition temperature being at $+80°C$ (compared with $-115°C$ for polythene). On subsequently heating to temperatures above 80°C it then begins to crystallize, and it is possible to explore the dependence of rate of crystallization on temperature. Such studies give the same kind of result as that obtained with rubber, the rate passing through a maximum at some intermediate temperature (fig. 7.9). In the neighbourhood of the

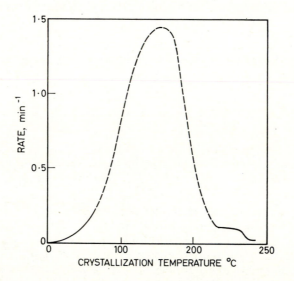

Fig. 7.9. Dependence of rate of crystallization of Terylene on temperature (Keller, Lester and Morgan, 1954). In the region shown dotted the rate of crystallization is too rapid to permit precise measurements.

108

maximum the curve cannot be followed precisely, but the two ends of the curve are clearly defined. It is significant that crystallization is limited to temperatures lying between the glass-transition temperature at the lower end, and the melting point at the upper end, just as in the case of rubber.

The results quoted above are typical of the crystallization behaviour of crystalline polymers. Taking into account all the available observations, we must conclude that the process of crystallization is essentially the same in all polymers which are capable of crystallizing, and that the rate of crystallization is governed by two factors, the rate of formation of nuclei and the mobility of the chain segments. The differences between different polymers are limited to such matters as the temperature range within which crystallization occurs, and the absolute value of the rate of crystallization. These features are related to the strength of the forces between the molecules, the ease of packing of the molecules in the unit cell, and so on, and vary widely from one material to another.

6. *The amount of crystallinity*

It has already been pointed out that for purely geometrical reasons it is not to be expected that 100% crystallization could be obtained in any polymer; there must always remain some fraction of the material in which the segments of molecules cannot be disentangled so as to be fitted into the lattice of one or other of the independently growing crystallites. In the case of rubber it was mentioned that the maximum amount of crystallization corresponds to a state in which only about 30% of the material is actually in the crystalline form. Is this a typical figure, or do polymers differ in the extent to which they crystallize?

Before attempting to answer this question some indication must be given of the methods available for determining the absolute amount of crystallinity in polymers. There are two principal methods, the density method and the X-ray method.

Density method. Let us imagine first that we could obtain the polymer (*a*) in the wholly amorphous state and (*b*) in the wholly (100%) crystalline state. In this imaginary case we could measure the density in both states and hence obtain the change in density corresponding to the transformation from the amorphous state to the fully crystalline state. Given a sample of the polymer which was only partially crystalline, a measurement of its density would then immediately give us the percentage of crystalline material present. If, for example, its density were just halfway between the figures for the amorphous and for the fully crystalline material, we would conclude that it was 50% crystalline.

This ideal procedure is impossible for the simple reason that though (in general) we can obtain the material in the completely amorphous state, we cannot obtain it in the fully crystalline state; we cannot therefore measure the density of the crystalline material directly. Fortu-

109

nately, however, there is an even better way of obtaining the density of the crystalline state, namely by calculation from the dimensions of the unit cell in the crystal lattice, as determined from the X-ray diffraction pattern. From these dimensions, which can be determined with considerable accuracy, we obtain the *volume* of the unit cell. The X-ray analysis also tells us the number of chains passing through the unit cell, and the number of monomer units or chain segments accommodated in the length of the unit cell. In the case of polythene, for example, the unit cell (fig. 7.3) contains two chains (one at the centre and a quarter at each corner) and includes two CH_2 groups in each chain; it therefore contains four CH_2 groups in all. Knowing the mass of each monomer unit we can thus obtain the mass of material in the unit cell, and hence the density of the crystal.

The type of result obtained is most simply illustrated by the case of natural rubber. The measured density of amorphous rubber is 910 kg m^{-3}, while the calculated density of the crystal is 1000 kg m^{-3}. There is thus an increase of density of 90 kg m^{-3}, or approximately 10%, on complete crystallization. The observed increase of density (or reduction of volume) on crystallizing rubber is only about 3%; the fraction of crystalline material is therefore only about 30%.

A slight difficulty arises in the case of polymers such as polythene which cannot be obtained in the completely amorphous state at room temperature. In this case it is necessary to calculate the density at room temperature from the density measured at a temperature above the crystal melting point together with the measured coefficient of volume expansion for the amorphous material. A more serious difficulty is encountered in polymers such as cellulose which cannot be melted; for such materials the method is not applicable.

X-ray method. Although the preceding method makes use of X-ray analysis, this is only for a ' once-for-all ' determination of the unit-cell dimensions. The X-ray method as used for the determination of the degree of crystallinity of a particular polymer sample involves intensity measurements on the X-ray diffraction photographs of the actual polymer to be studied. The precise methods vary according to the material, but they depend, in general, on a separation of the sharp spots, or rings (as the case may be) associated with the crystalline reflections from the diffuse halo or ' background ' scattering due to the amorphous component (Plate 1 (c)). As the amount of crystalline material present is increased the intensity of the crystalline reflections increases, while the intensity of the amorphous halo diminishes, hence by comparing intensities the degree of crystallinity may be estimated.

Results. The degree of crystallinity of a polymer varies to some extent according to its method of production, and also according to the mechanical treatment (e.g. drawing) or thermal treatment to which it has been subjected. These mechanical and thermal treatments may affect the processes of crystallization. In addition to such genuine differences

110

the figures obtained for the degree of crystallinity may vary according to the method of estimation applied; different methods do not always yield consistent results. The figures must therefore be regarded as only approximate, and may be subject to an uncertainty amounting perhaps to about 10% of the total material.

ESTIMATED DEGREE OF CRYSTALLINITY (%)

Natural rubber (crystalline)	20–30
Gutta-percha	55–60
Polythene (low-density)	55–75
Polythene (high-density)	80–95
Nylon (fibre)	50–60
Terylene (fibre)	60
Cellulose (cotton, flax)	70
Cellulose (viscose rayon)	40
Polypropylene (fibre)	55–60

This table shows that the crystalline polymers generally have crystallinities in excess of 50%. Cellulose is peculiar; the natural cellulosic fibres are highly crystalline, but in man-made fibres of cellulose the degree of crystallinity is rather low. The difference between high-density and low-density polythene is also significant; in high-density polythene there are no branches or side-chains to interfere with the crystallization, and in this material the crystallinity may be surprisingly close to the theoretical limit (100%). Its crystallinity is very much higher than that of any other known polymer. High-density polythene is noticeably harder and less flexible than the low-density form; it also has a higher melting-point.

7. *Polymer single crystals*

In the discussion of the processes of crystallization we have emphasized the virtual impossibility of obtaining 100% crystallization in a polymer, and it has been tacitly implied that anything resembling a single polymer crystal of visible dimensions is an unattainable ideal. One of the great surprises of the last ten years or so has been the discovery that under certain very special conditions it is possible to obtain single crystals of polyethylene, polypropylene and a number of other polymers of sufficiently large size to be easily visible in the electron microscope (Plate 2). Since their discovery these single crystals have been the subject of intensive study, particularly by Keller in Bristol and Geil in U.S.A., and the facts elucidated concerning their structure and properties have been almost as surprising as the fact of their existence itself.

Single crystals of polymers are obtained by slow precipitation from very dilute solution. They have the form of extremely thin plates or

111

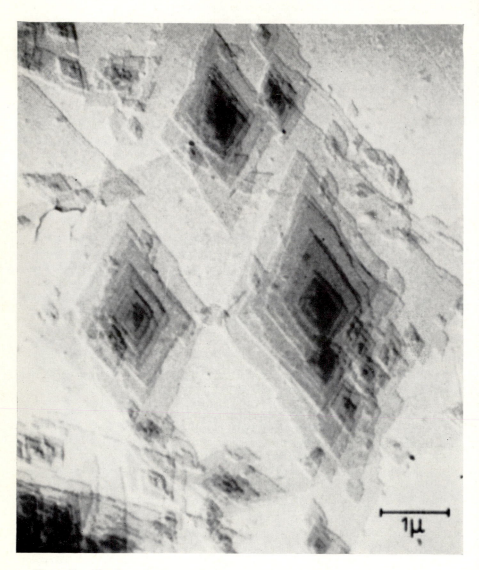

1μ

PLATE 3.   Single crystals of polyethylene as seen in the electron microscope.
(Magnification 18 000 ×) (A. Keller).

112

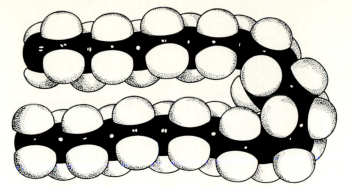

Fig. 7.10. Formation of fold in polyethylene chain.

lamellae (Latin *lamella*, plate) which in the case of polyethylene are of rhomboidal or lozenge shape. Spiral growth patterns are usually visible on their surfaces; these correspond to a step-like formation on the surface, the height of each step being usually between 10 and 20 nm (100 to 200 ångström). The dimensions in the plane of the lamellae may be quite large, up to 0·01 mm or more.

The structure of these single crystals is remarkable. The reader will note that the thickness of the lamellae is very small compared with the length of the polymer chain. It would therefore seem natural to suppose that the molecular chains are arranged with their axes lying in the plane of the lamellae. Very surprisingly, electron diffraction studies have established beyond any doubt that the chain direction is *perpendicular* to the flat surfaces of the lamellae, in a direction, that is, which can accommodate only about one hundred chain carbon atoms out of a total of some 1000 or 10 000 in the whole chain. This raises an acute problem concerning the mechanism of formation of the single crystal.

There seems to be only one solution: the chains must somehow or other be bent or folded backwards and forwards across the thickness of the lamellae.

Further examination of the properties of the polymer chain suggest that this conclusion may be less difficult to accept than might at first sight be imagined. The flexibility of the chain, which is associated with rotations about single bonds, is such that in the case of polythene, for example, a fold may be produced, with very little increase in bulk, in as few as three successive C–C bonds (fig. 7.10). Such folded chains can be stacked in the crystal lattice without too much difficulty. The generally-held opinion is that the single crystal consists of a regular array of folded chains laid down individually and successively on the growing edges of the lamella (fig. 7.11).

This type of crystallization is made possible by the comparatively large distances which separate the individual molecules in the dilute solution

113

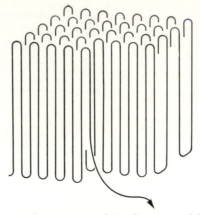

Fig. 7.11.   Diagrammatic representation of process of formation of folded-
chain single crystal.

(cf. fig. 2.3 (*a*), p. 20).   As a result of this wide separation the entanglements between chains which normally restrict the freedom of motion of
chain segments are no longer present.   Under these conditions it is
apparent that if one segment of a polymer molecule becomes attached to
one of the thin edges or steps of the growing crystal, there will be no competition from other molecules for the occupation of neighbouring lattice
sites and nothing to prevent the successive deposition of segments of the
same molecule on the immediately adjacent sites.   This deposition
evidently takes place by the folding up and down of successive segments
of the chain until the whole molecule is attached.   This folding may not
be completely regular in the first instance (there may be irregular loops
projecting from the plane surface) but these irregularities can be subsequently adjusted by slippage of chains within the crystal or movement
of chain folds along the length of the chain.

8. *Structure of crystalline polymers*

Although single crystal lamellae have only been isolated when crystallization takes place from dilute solution, their discovery has had a profound effect on our ideas concerning the structure of crystalline polymers
in bulk as obtained on cooling from the molten state.   It would be
surprising if a form of crystallization which appears to be so strongly
favoured in one particular type of environment did not have some part to
play in a somewhat different environment, and the view is now widely
held that the lamella is the basic unit of structure in bulk polymers also.
There is considerable evidence to support this view.   The fractured
surfaces of crystalline polymers, when examined under the electron
microscope, frequently reveal a series of uniformly spaced parallel planes
strongly suggestive of lamellae stacked one on top of another, and having
spacings of the same order of magnitude as the single crystal lamellar

114

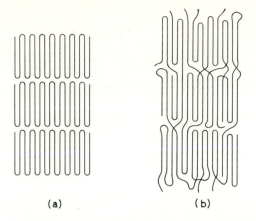

(a)   (b)

Fig. 7.12. (a) Idealized representation of stacking of lamellar crystals. (b) Modified form of (a) showing various types of imperfection.

thickness (fig. 7.12 (a)). X-rays also show reflections at very small angles which represent some fairly regular periodicity in the structure, corresponding to a dimension equal to the thickness of the lamellae. Such a periodicity would arise from the regularly spaced gaps, or regions of low density, between successive lamellae.

Nevertheless there are difficulties in accepting the lamellar structure in the extreme form depicted in fig. 7.12 (a) as a model which is applicable to all types of polymer. Except for high-density polythene, most polymers contain a very considerable fraction of material, varying from about 25% to 50%, which is not in the crystalline state. A place has to be found for this amorphous material. In the type of model which was discussed in connection with crystalline rubber, which may be called the 'fringed-crystallite' model (fig. 6.4 (a)), the amorphous material is assumed to occupy the regions between the randomly oriented crystallites. This model is probably nearer to the truth for a polymer like rubber, which is not very highly crystalline. For more highly crystalline polymers various intermediate types of structure between the fringed-crystallite model and the model of regularly stacked lamellae of the type represented in fig. 7.12 (a) may be envisaged; these allow for varying degrees of imperfection corresponding to such features as irregular fold lengths, protruding loops of varying length, interconnecting chains between lamellae, and some degree of entanglement in the regions between lamellae, fig. 7.12 (b).

Looked at in this way, there is no logical contradiction between the fringed crystallite model and the lamellar model of the structure of crystalline polymers. These two types of structure may be regarded as the two theoretical extremes, neither of which is ever actually realized. Actual polymers lie somewhere between these two extremes. The very

115

highly crystalline polymers, notably high-density polythene, may approach very closely to the idealized lamellar type of structure. For weakly crystalline materials, such as rubber, and possibly viscose rayon, the fringed-crystallite model probably represents a closer approximation to the structure. Other polymers may have some of the features of each of these simplified types. This is a matter which can only be decided by a careful study of the particular polymer, using all the available methods of examination of structure (X-ray diffraction, electron microscopy, etc.).

CHAPTER 8

fibres and fibre formation

1. *Introduction*

IN Chapter 1 some consideration was given to the general structural characteristics of fibres. Fibres are polymers which are characterized by a high degree of molecular orientation or alignment; they are also usually highly crystalline. The structure of a fibre is, in fact, very similar to that of a highly extended crystalline rubber (fig. 6.4 (*b*)). It is the high degree of molecular alignment which enables the fibres to withstand high tensile stresses without suffering unduly large deformations. Much of the art of fibre production is concerned with the precise means by which the desired orientation of the molecules or crystallites is brought about; this varies according to the nature of the polymer and the type of process employed in the production of the fibre.

In the present chapter we shall examine some of the typical polymers used in fibre production and the processes by which they are converted from the original raw state, corresponding to the unoriented material in bulk, into the final state consisting of highly oriented very fine filaments.

2. *Nylon*

Although we are accustomed to thinking of nylon as a single material, the term *nylon* actually covers a whole range of polymers, all of which have the same basic type of structure. These polymers are known as *polyamides*—an amide being a compound containing the amide grouping $-CONH_2-$. The most commonly used form of nylon in Great Britain has the very long repeating unit, containing 14 atoms in the chain backbone structure, shown in fig. 8.1. It will be seen that this unit consists of two sequences of six carbon atoms each, terminated by the single nitrogen atoms of the amide groups; its technical designation is nylon 66 (read as *six-six*). Another nylon, known as nylon 610 (six-ten), has the even longer repeating unit of eighteen atoms, this being made up of alternate sequences of six and ten C atoms similarly separated by single N atoms.

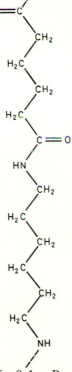

Fig. 8.1. Repeating unit in nylon 66 chain.

The zig-zag structure of the nylon chain is very similar geometrically to that of polyethylene, and may be regarded as a polyethylene chain interrupted by occasional C=O and N–H groups. The presence of these groups, however, has a very important effect on the forces between the molecules and on the chemical properties of the polymer. One very striking consequence of this is seen in the structure of the crystal, in which the parallel chains line up in such a way that the C=O group of one chain and the N–H group of the neighbouring chain come together (fig. 8.2); this pattern of association

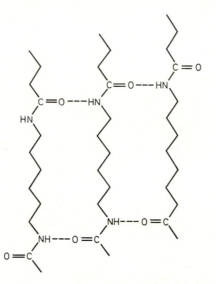

Fig. 8.2. Arrangement of neighbouring chains in crystal of nylon 66.

is repeated along the whole length of the chain and throughout the crystal. The reason for this is that these two groups have a very special attraction for each other, an attraction amounting almost to a type of chemical bonding. A partial bonding of this kind is usually indicated in the following way.

$$C=O- - - -H—N. \tag{8.1}$$

In this partially bonded state the O and N atoms are brought much closer together than they would otherwise be, the H atom acting as a kind of bridge between them, forming what is called a *hydrogen bond*. The high melting point of nylon (265°C) is the result of the great stability of the crystal structure resulting from the strong hydrogen bonds within the crystal. In polythene there are no such special attractive forces between the chains and the melting point is consequently very much lower (130°C).

118

3. Terylene

Terylene is a *polyester* fibre. Though polyesters of many types occur in the plastics industry, as well as in paints and adhesives, Terylene is the only polyester fibre of major importance at the present time.

An ester is the product of the reaction between an acid and an alcohol. The name 'Terylene' is derived from the acid constituent, which in this polymer has the rather awkward designation *terephthalic acid*, and is represented by the formula

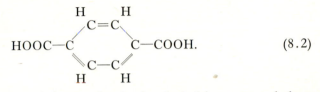

$$\text{HOOC—C} \quad \quad \text{C—COOH.} \qquad (8.2)$$

The other component of the reaction is the alcohol known as ethylene glycol, i.e.

$$\text{HO—CH}_2\text{—CH}_2\text{—OH.} \qquad (8.3)$$

In the polymerization reaction the terminal groups of (8.2) and (8.3) link up (with the loss of one molecule of H_2O) to produce a regular sequence of alternate (8.2) and (8.3) groups, whose chemical name is *polyethylene terephthalate**. The chain repeating unit has the formula

$$\text{—CO—C} \quad \quad \text{C—CO—O—CH}_2\text{—CH}_2\text{—O—} \qquad (8.4)$$

The Terylene chain is of a very different geometrical form from the simple zig-zag structures of polyethylene and nylon. This form is shown in fig. 8.3, in which, in order to show more clearly the symmetry of the structure, the benzene ring has been taken as the centre of the repeating unit. This central ring, together with its two associated C–C bonds, forms a single rigid structure which does not have the possibility of changing its configuration in any way. Rotational movements within the chain are limited to the C–O and C–C single bonds. As a result the chain is less flexible or rubberlike than the polythene or nylon chains. This, together with its rather 'bulky' character, could account for the slower rate of crystallization of Terylene, compared with polythene and nylon. As a result of this slower rate of crystallization, Terylene may be cooled from the melt without appreciable crystallization, whereas in the case of both polyethylene and nylon this is not possible (cf. Chapter 7). The high glass-transition temperature of the amorphous Terylene, 80°C, is probably also connected with these same structural peculiarities. The

---

* In U.S.A. this material is marketed as 'Dacron'.

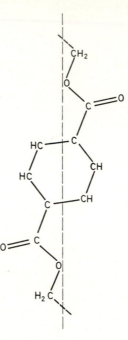

Fig. 8.3.   Repeating unit in Terylene chain (chain axis shown dotted).

melting temperature of Terylene is, however, similar to that of nylon 66, namely 265°C.

4. *Melt spinning*

The process of production of a fibre from the bulk polymer is known as *spinning**, and may be carried out either from the melt or from solution. Polymers such as nylon, Terylene and polypropylene, which can be readily melted, are usually melt-spun. The melt-spinning process consists in the extrusion of the molten polymer at a carefully controlled rate through a number of fine holes in a die or *spinneret*. The filaments issuing from the die cool down and solidify in their passage towards the winding-up gear. The general arrangement is illustrated in fig. 8.4. The polymer, in the form of small pellets or *chip*, is fed from the hopper into a chamber from which air is excluded. In this chamber it is melted by means of the electrically heated ' melt grid ' and maintained at the required temperature before being pumped through the spinneret. The tension in the ' threadline ' is comparatively light, and is not sufficient to

---

* The original meaning of this term was to *stretch out* or *span*, as a spider *spins* its web. By association it is applied also to the twisting process by which a yarn is formed, this being combined (in the case of short fibres such as cotton) with an extension or drawing process.

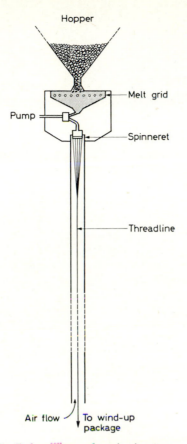

Fig. 8.4. The melt-spinning process.

produce appreciable stretching of the filaments after they have solidified. In this state, therefore, the polymer, although in filament form, is still without molecular orientation or alignment.

5. Drawing and ' necking '

The undrawn polymer is capable of being extended to several times its original length when a suitable stretching force is applied. This high extension, however, is not elastic and reversible, as in vulcanized rubber, but irreversible or inelastic; the stretched polymer does not return to its original length when the stretching force is removed.

The relation between the extension and the applied force has the form shown in fig. 8.5. The first portion of the curve, up to perhaps 10% extension, is approximately linear. In this region the deformation is more or less reversible. The force then reaches a maximum and sub-

121

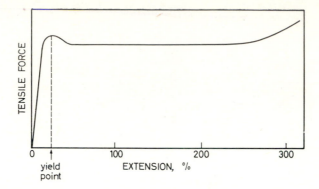

Fig. 8.5. Force–extension relation for undrawn crystalline polymer.

sequently falls slightly; the extension at which the maximum occurs is called the *yield point*, and the corresponding stress the *yield stress*. At this point the structure in some way breaks down or yields; thereafter the deformation is not recoverable. The stress subsequently remains constant as the strain increases until the ultimate breaking point is approached, when it again begins to rise.

The type of deformation which occurs on drawing may be more conveniently studied by using test-pieces cut from a sheet of bulk polymer, rather than the polymer in filament form. Examination of the process of extension reveals some remarkable effects. Up to the yield point the extension is perfectly uniform; all parts of the specimen are equally strained. At the yield point, however, a thinning or 'necking' occurs at one particular point (fig. 8.6 (*b*)), and spreads until a region of reduced cross-sectional area can be distinguished from the remainder of the specimen (*c*). As the extension proceeds the length of the region of

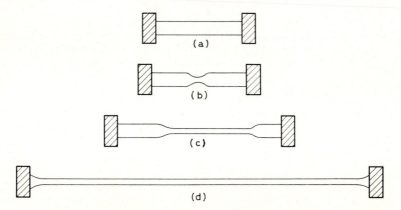

Fig. 8.6. Intermediate stages in the drawing of a crystalline polymer, showing neck formation.

122

reduced area increases at the expense of the remaining portion until the whole length of the specimen, up to the neighbourhood of the clamps, has become uniformly stretched (*d*). This final stage is then followed by rupture.

The above experiment can be easily performed on a strip of polythene, say 10 cm × 1 cm, cut from ordinary commercial sheet. In the course of the extension the specimen is divided into two distinct segments, one highly drawn, and the other substantially undrawn. Further extension merely increases the *amount* of material in the drawn state, without affecting the state of strain of the already drawn material. This is why the tension remains constant during the drawing process.

The industrial drawing process is essentially the same as that described, but is carried out continuously on long lengths of yarn instead of on a short specimen. This is achieved by passing the yarn through two pairs of rollers, the front pair, A, being rotated at a faster rate than the rear pair B (fig. 8.7). If nothing else were provided the ' neck ' would move

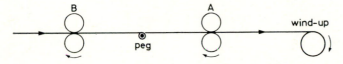

Fig. 8.7.   Continuous drawing of yarn.

about spasmodically between A and B, producing irregularities in the yarn. In the case of nylon this is avoided by looping the yarn round a peg P; the friction thus developed modifies the stress and so stabilizes the position of the neck. In the case of Terylene satisfactory drawing can only be carried out when the polymer is heated to a temperature above the glass-transition temperature. This is done by passing the yarn through a heated zone, and this automatically stabilizes the position of the neck.

The characteristics of the drawn polymer can be varied to a certain extent by varying the conditions of drawing, particularly the rate of extension and the amount of extension or draw ratio. Fig. 8.8 shows typical stress-strain curves for high-tenacity (i.e. high strength) and medium-tenacity nylon yarns. The high-tenacity yarns are of the type used in tyre cord, where the maximum tensile strength is required, but the high tensile strength is coupled with a high modulus of elasticity and low extensibility. The medium-tenacity yarn has a lower strength, but also a lower modulus and higher extensibility—properties which are desirable in knitted and woven fabrics for clothing uses, where softness and flexibility are more important than tensile strength.

6. *Cellulose and its derivatives*

Although the synthetic fibres are being produced in increasing quantities and in a great variety of types, the largest proportion of the

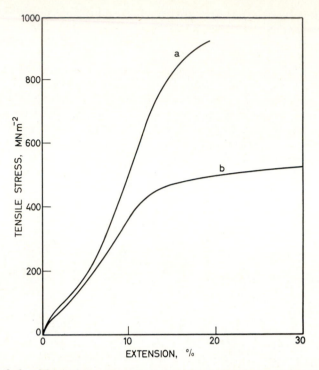

Fig. 8.8.   Typical stress–strain curves for (*a*) high-tenacity and (*b*) medium-tenacity nylon yarns (Hall, 1964).

man-made fibres appearing on the world's markets is still based on cellulose.   The most important of these is viscose rayon, which is a pure cellulose derived mainly from wood-pulp.   The other cellulosic fibres comprise the chemically modified forms of cellulose, notably cellulose acetate and cellulose triacetate (tricel).   Bearing in mind the additional consumption of cellulose in the form of the natural fibres—cotton, flax, jute, etc., we obtain some idea of the dominant position which cellulose still occupies in the fibres industry.

In addition to its outstanding importance in the industrial world, cellulose, as was already noted in Chapter 1, holds a pre-eminent position in the biological field, as the almost universal basic component of plant structures.   Being also one of the first of the polymers to be extensively used in the chemical industry, it is not surprising to find that cellulose has been subjected to a more intensive scientific study from all points of view—chemical, physical and biological—than any other of the fibre-forming polymers.   Such a large amount of study has indeed been necessary, for the cellulose molecule is a peculiar and rather complicated structure, very much more complicated than the molecules of the com-

124

mon synthetic fibres already considered, and its capacity for undergoing chemical change is correspondingly complex.

7. *Structure and properties of cellulose*

The cellulose chain is very closely related to glucose, which is one of the simpler sugars. The glucose molecule has the formula $C_6H_{12}O_6$, and consists of a ring containing five carbon atoms and one oxygen atom, together with various attached groups. The glucose ring is not flat (like the benzene ring) but crumpled or 'staggered', as indicated in fig. 8.9 (*a*). The cellulose molecule may be regarded as being formed by the linking together of a succession of glucose rings at their terminal OH groups, one molecule of water being removed at each such point of combination. The unit of the chain structure is therefore the so-called *anhydroglucose* unit (i.e. glucose *without* water) (fig. 8.9 (*f*)). The

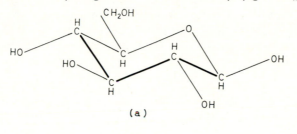

(a)

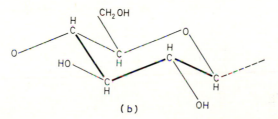

(b)

Fig. 8.9. Structure of (*a*) glucose ring, (*b*) anhydroglucose unit.

skeleton structure of the chain has the form shown in fig. 8.10. It is seen that the successive anhydroglucose units are not in geometrically equivalent positions but are alternately rotated through 180° about the chain axis. Thus, although the chemical repeating unit is the single ring, the geometrical or structural repeating unit consists of *two* anhydroglucose units.

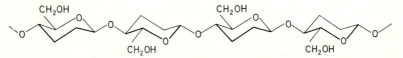
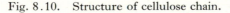

Fig. 8.10. Structure of cellulose chain.

125

The three OH-groups attached to each of the rings in the chain have an important influence on the properties of the cellulose molecule. First, they have a strong attraction for water molecules (H–O–H), to which they bear a close chemical resemblance, with the result that cellulose fibres are capable of absorbing a considerable amount of water. Such absorption of water is accompanied by a swelling of the fibre—a subject dealt with more fully in Chapter 10. Secondly, these same OH-groups exert a strong attractive force on the OH-groups in neighbouring chains, forming ' hydrogen bonds ' between molecules of a similar type to those discussed in connection with nylon. Owing to the large number of these bonding units (three for each glucose ring), the total attractive force between chains is, however, very much higher than in the case of nylon. As a result the cellulose crystal is extremely stable, the crystal melting point being above the temperature of decomposition. Cellulose cannot therefore be melted, nor can it be dissolved in water or any of the ordinary solvents. Thirdly, by suitable chemical treatment it is possible to replace each or all of the three OH-groups by a different group; in this way a number of chemically modified forms of cellulose of considerable importance may be obtained. The replacement of the OH-groups leads to a reduction in the strength of the intermolecular forces, and, in the case of partial substitution particularly, to a reduction of crystallinity. In consequence such modified forms of cellulose are generally soluble and may therefore be more readily extruded or otherwise handled. One of these modified celluloses is cellulose nitrate, which in the form of *celluloid* was one of the earliest of the ' plastics ' in general use. Celluloid, however, has the great disadvantage of being almost explosively inflammable. The fully nitrated form of cellulose, containing three nitrate groups on each glucose ring (tri-nitro-cellulose) is in fact the explosive known as gun-cotton, and is a constituent of the present-day cordite. For this reason another modification, cellulose acetate, which does not suffer from this disadvantage, has almost entirely replaced the nitrate in all important applications, e.g. toys, photographic films, etc.

8. *Spinning of cellulose*

Owing to its insolubility in ordinary solvents the processes used in producing cellulose fibres are of a very special kind. There are various ways of getting cellulose into solution, but they all involve some chemical modification of the original structure, followed by a further chemical treatment after formation of the filaments to convert them back into cellulose. One of the simplest of these processes, from the chemical standpoint, is the acetate process, in which the spinning is actually carried out on cellulose acetate in an acetone solution. This is an example of the process known as *dry spinning*; after extrusion the filaments are passed through a stream of hot air which dries off the solvent. The acetate is then converted back to cellulose by suitable chemical treatment or

126

hydrolysis. More commonly, regenerated cellulose is produced by the process known as *wet spinning*. In this the solution of the chemically modified cellulose is extruded directly into a liquid bath in which the filaments are converted back to cellulose while still in the liquid state. One such process, the *cuprammonium rayon* process, involves solution, with chemical modification, in a medium containing a copper salt plus ammonia. The process most widely used today, however, is the *viscose rayon* process, invented by Cross and Bevan in 1892. In this the cellulose is converted to a sulphur-containing compound known as cellulose xanthate, the xanthate filaments being converted back to cellulose in a bath containing sulphuric acid and certain salts.

9. *Properties of cellulose filaments*

As with the melt-spun polymers, the mechanical properties of the final filament depend very much on the amount of molecular orientation introduced. But whereas in melt-spun polymers this orientation is produced by a separate and distinct drawing operation, in the case of cellulose the processes of orientation and formation of the final structure are intimately bound up with the actual chemical regeneration process itself in a rather complicated way. To understand how this is brought about it is necessary to take a closer look at the regeneration process. The viscose solution is extruded through the spinneret directly into the acid bath. When it enters the bath two effects occur. First, the cellulose xanthate is immediately precipitated from solution to form a milky-looking highly-swollen solid of jelly-like consistency containing a large proportion of water. Secondly, the coagulated xanthate is acted on by the sulphuric acid, which rather slowly converts it into cellulose. During this process the water diffuses outwards from within the filaments and they become progressively harder and stronger with increasing conversion to cellulose. At the same time the filaments are being pulled out and wound up at a faster rate than they are being extruded from the spinneret, so that there is both a lateral contraction and an axial extension, which together produce a moderately high degree of molecular orientation. For the production of high-strength yarns, as used for example in tyre cord, an additional extension process may be carried out at the stage where the chemical conversion is complete, though the filaments still contain a considerable amount of water.

The process of crystallization takes place simultaneously with the formation of the cellulose, and with the removal of water by diffusion outwards. The orientation of the crystallites reflects the prevailing orientation of the molecules at the time of their formation, though it may be further enhanced by subsequent axial extension of the filament, and by lateral contraction on drying. Once the filaments are completely dried it is not possible to change the state of orientation by further drawing, as is done in the case of melt-spun polymers.

127

The differences in mechanical properties associated with differences in degree of orientation are comparable to those observed in melt-spun polymers. Increased orientation gives an increase in strength, with a corresponding reduction in extensibility. Figure 8.11 shows typical stress–strain curves for a textile rayon, and for a tyre-cord rayon. These may be compared with the natural fibres, flax and cotton, which possess not only a more perfect orientation but also a higher degree of crystallinity than can be produced in regenerated fibres (Chapter 7).

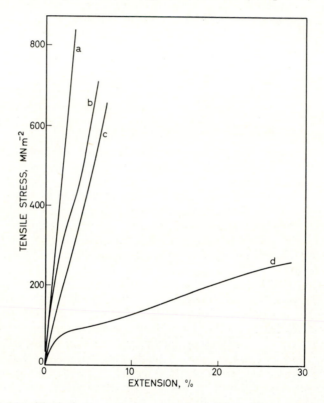

Fig. 8.11. Typical stress–strain curves for (a) flax, (b) tyre-cord viscose rayon, (c) cotton, (d) textile viscose rayon (Meredith, 1956).

10. Cross-section of filaments

Whereas melt-spun filaments are usually approximately circular in cross-section, as shown in fig. 8.12 (a), filaments of viscose have the very curious and irregular form of cross-section shown in fig. 8.12 (c). This arises from the process of formation in the spinning bath. On coming into contact with the acid, the outermost layer of the filament is converted into cellulose almost immediately. This forms a skin on the

128

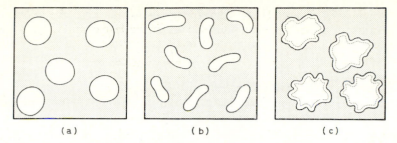

| (a) | (b) | (c) |

Fig. 8.12. Typical forms of filament cross-section. (a) Nylon. (b) Orlon.
(c) Viscose rayon.

surface, within which the original xanthate is contained in a highly
swollen state. Subsequent reaction can take place only by diffusion of
acid through this skin. As the conversion of the inner layers proceeds
water is extruded by diffusion outwards and the filament collapses to a
smaller volume. This shrinkage leads to wrinkling of the skin, which
gives rise to the folded or corrugated profile shown in fig. 8.12 (c). The
presence of the skin can readily be detected by its different rate of dye
uptake.

This form of profile, though not produced deliberately, has certain
practical advantages. It affects the packing of the filaments in the yarn,
leading to a more open, less dense, yarn which gives a softer and more
desirable feel in textile fabrics. It also affects the reflection of light,
giving a whiter and less shiny or lustrous appearance which is generally
more desirable. In melt-spun filaments it is frequently necessary to
incorporate a 'delustrant' in the polymer melt in the form of a white
powder in order to counteract the excessive gloss.

11. *The acrylic fibres*

The acrylic fibres—or acrylics, as they are usually called—are an
important class of fibres, marketed under the trade-names Courtelle,
Orlon, Acrilan, etc. The main constituent of the acrylic fibres is the
polymer known as polyacrylonitrile, based on the acrylonitrile unit

$$-CH_2-CH- \atop | \atop CN \qquad (8.5)$$

but in addition to this main constituent, there is a smaller amount of
some other unit, which may be one of a number of different monomers.
It will be seen that the basic polymer has a similar form of structure to
polypropylene and polyvinyl chloride, i.e.

$$-CH_2-CH- \atop | \atop X \qquad (8.6)$$

129

in which X is either CH_3 or Cl. Materials of this kind are known as *vinyl* polymers, and, as was pointed out in the last chapter, they have various possibilities of arrangement of the side-group X with respect to the central carbon atom. Polyacrylonitrile is an *atactic* polymer; it therefore does not crystallize. It is an exception to the general rule that polymers which form fibres should be capable of crystallization. Actually, the important factor is not crystallization as such but molecular orientation or alignment, which is usually facilitated by crystallization. This has the effect of strengthening the forces between the molecules and stabilizing the oriented structure. In polyacrylonitrile the forces between the chains arising from the presence of the CN groups are already rather strong, with the result that the oriented state can be maintained despite the absence of crystallization. (Of course, if an *isotactic* polyacrylonitrile were produced, this would be expected to be crystalline *and* fibre-forming.)

The acrylic fibres are not melt-spun, for the simple reason that they cannot be melted. Not being crystalline, they do not have a *crystal* melting point, but whereas most polymers ultimately soften and flow at higher temperatures, the softening point of the acrylics is above their decomposition temperature. They can, however, be dissolved in certain solvents and extruded in solution form, the original polymer being recovered from solution either by evaporation of the solvent (dry-spinning) or by coagulation in a suitable liquid bath (wet-spinning). The choice of process is dependent on the nature of the second ingredient of the copolymer. Subsequent to extrusion, and while still containing some liquid, the filaments are subjected to a stretching operation to produce the required molecular orientation. The form of cross-section varies with the method of production but is typically flattened due to collapse on drying. This is illustrated in fig. 8.12 (*b*), representing the cross-section of filaments of Orlon.

12. *Conclusions*

The range of materials from which fibres may be formed is very extensive, and in the present chapter we have only been able to pick out a few of the more important of these materials for special consideration. Our purpose has been to illustrate the principles involved in the production of a fibre, and the way in which the final properties are achieved.

The variety of materials which are capable of being produced in fibre form brings out the important point that it is mainly the *physical* properties of the polymer which determine its suitability for any particular use. In the case of fibres, the important physical properties are high strength, high modulus of elasticity or resistance to deformation, and high softening temperature. These properties are dependent not only on the basic chemical nature of the polymer, but also in a very special way on its physical state or structure, i.e. on the crystallinity, degree of orientation, etc., and this, in turn, is largely determined by the technical operations by which it is produced. It is possible to alter the physical properties within wide limits by suitable modifications of the processes of production. It follows from this that it is extremely difficult, *in advance*, to forecast whether a given polymer will be capable of forming a commercially successful fibre; so much depends on the technical developments taking place all the time in the production and utilization of fibres and on the specific market needs which it may satisfy, and also on the economic factors involved, i.e. cost of raw materials, scale of production, competition from other sources, etc. It requires many years of effort—and a sum of money amounting probably to tens of millions of pounds—to develop a marketable fibre, and not all such developments turn out to be ultimately profitable. But if the penalty for failure is high, so also is the reward for success, and the economic return on a successful fibre can be very considerable.

CHAPTER 9

strength and fracture

1. *Significance of strength*

MATERIALS are used for any particular purpose primarily because they possess certain physical properties which are essential to that purpose. They must, of course, also satisfy certain chemical requirements, but for most everyday uses these are not unduly stringent and can usually be met without too much difficulty. Broadly, it is the physical properties of the material which are the more critical and which set the limits to its practical applications. If any particular material possesses the right physical properties to satisfy the requirements of a specific practical situation— and these include satisfactory performance during the necessary processes of fabrication as well in the final utilization of the product—then this material will be used, regardless of whether it is natural or synthetic, organic or inorganic, polymeric or metallic. Polymeric materials thus find themselves in competition with a great variety of materials of all kinds, and if they are to advance their position in the industrial world the full range of their potential physical properties must be exploited in the best possible manner. This requires the acquisition of systematic scientific knowledge concerning their physical properties and of the way in which these physical properties may be controlled and put to the most effective use.

Foremost among the physical properties of practical importance is the property of *strength*. However attractive in other respects, if a material fractures or breaks down physically in service it is a failure. This is true whether the article is a railway bridge or a refrigerator, the wing of an aeroplane or the windscreen of a car. Plastics in particular have acquired an unfortunate reputation for brittleness in certain applications, sometimes because they were unsuitable for the purpose, but more often because the article has been designed without proper attention being paid to the stresses likely to be encountered in actual use.

The subject of strength is a peculiarly difficult one, for a variety of reasons. Strength is not a physical property in the same sense as (to take just a few examples) the density, the hardness or elastic modulus, or the electrical conductivity of a material. If each of a dozen schoolboys is asked to find the density of glass, given similar specimens, their results will probably all agree to within a few parts in a thousand. But if they are each given a similar glass rod and asked to measure its strength, for example in bending, individual results will show large

132

variations from the average value, amounting to perhaps ± 50% or even more. These variations are not due to errors of measurement as usually understood, but are inherent in the nature of the property which is being measured. A typical test result for sheet glass is represented in fig. 9.1, which shows the numbers of specimens having values of strength within any given range. From this it is seen that even under the most favourable laboratory conditions apparently identical specimens of the same material may differ very considerably in strength.

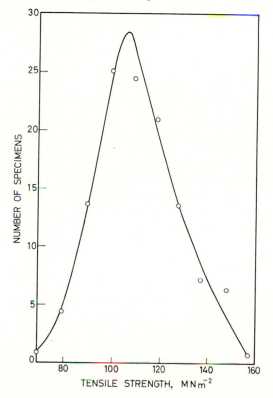

Fig. 9.1. Distribution of measured values of strength of sheet glass (Holland and Turner, 1934).

This inherent variability is due to the fact that strength is what is called a *structure-sensitive* property. By this we mean that any small variations or irregularities in the structure of the material, which would have no significant effect on other physical properties, can have a profound and quite disproportionate effect on its strength. These variations may arise in many ways, for example, through minor differences in composition from point to point in the material, from the presence of particles of dust or other impurities, or from stresses arising from mechanical treatment

133

during processing. By far the most important, however (as we shall see later), are irregularities due to surface defects, e.g. very fine scratches or minor abrasions, such as are inevitably produced in the course of handling or processing.

Most of the commonly-measured properties of materials—density, heat capacity, refractive index, modulus of elasticity, etc., are average properties over the whole sample; any small variations in structure produce correspondingly small variations in such average properties. Strength, on the other hand, is an *extreme* property; failure occurs at the *weakest* point in the specimen. The well-known saying ' the strength of a chain is its weakest link ' embodies an important scientific principle which is involved in all measurements of strength. It is the point of exceptional weakness, the most serious flaw or defect in the structure, which determines the maximum stress which can be borne. By the very nature of these irregularities, this maximum strength must therefore vary widely from one specimen to another.

This kind of variability poses a difficult problem for the design engineer. Design must be based on the *minimum* value of strength which is ever likely to be encountered in practice. This minimum will certainly be much less than the *average* value, but it cannot be precisely specified. The practice has therefore grown up of incorporating into the design a ' factor of safety ' to cover this uncertainty. Factors of safety of ten or more are not uncommon in engineering practice. This may not be a great disadvantage in heavy engineering, e.g. in bridge building, but it may be an extremely serious factor in the design of an aeroplane. There is therefore an urgent need not only for the development of materials of high strength, but also for the utmost control of the factors on which the variability of strength depends.

2. *Typical values of strength*

Representative figures for the tensile strength of a number of polymeric and other materials are shown in the table below. For rubbery and glassy polymers the strength is usually below 100 MN m^{-2}. The effect of orientation of the molecules is to increase the strength; in the case of a glassy polymer, this increase may amount to a factor of about 2·0. The strength of fibres is also related to their molecular orientation, and the most highly oriented fibres, flax, tyre-cord rayon and high-tenacity nylon (also used for tyre cord) have values of strength in the range 700–1000 MN m^{-2}. Less highly oriented fibres, such as wool and ordinary viscose rayon, yield somewhat lower figures, which are, however, still well above that for an unoriented crystalline polymer such as undrawn nylon.

Comparing the polymeric fibres with metals, it is seen that they are stronger than aluminium but less strong than the highest-strength steel (piano wire). The difference between a high-tenacity tyre cord (rayon

134

Material	MN m^{-2}	kgf/mm^2
Iron, ' whiskers '	14000	1400
Steel, piano wire	2000	200
Mild steel	460	46
Carbon filaments	3200	320
Aluminium	170	17
Flax (cellulose)	900	90
Cotton (cellulose)	500	50
Viscose rayon, tyre cord (cellulose)	1000	100
Viscose rayon, textile fibre	300	30
Nylon, high-tenacity	1000	100
Nylon, undrawn	80	8
Wool	200	20
Glass, sheet	40–80	4–8
Glass, filaments	1000–3000	100–300
Perspex	80	8
Perspex, oriented	160	16
Urea-formaldehyde resins	40–80	4–8
Rubber	30	3

Table 9.1.   Tensile strengths of various materials.

or nylon) and steel, however, is not very great, amounting to a factor of about two only.

The figures given in the above table represent the strength calculated on the basis of the cross-sectional area of the specimen, i.e. the force divided by the cross-sectional area.   For many purposes, however (e.g. in aircraft), it is the *mass* of material required to support a given load which is the significant consideration.   This may be taken into account by comparing the strengths on a *mass* basis, for which purpose we divide the strength as defined above by the density of the material (relative to water).   On this basis the high-strength fibres compare favourably with steel, as the following table shows.   This table also includes graphite filaments, which will be discussed later in connection with the reinforcement of polymers.

The following examples may help the reader to visualize the significance of some of the above data.   A high-tenacity nylon filament of cross-sectional area 1 mm^2 (diameter 1·13 mm) would support the

Material	Strength T (MN m^{-2})	Relative density, ρ	T/ρ (MN m^{-2})
Steel	2000	7·8	260
Tyre cord rayon	1000	1·56	640
High-tenacity nylon	1000	1·15	870
Carbon filaments	3200	1·9	1700

Table 9.2.   Strengths calculated on an area (col. 2) and on a mass (col. 4) basis.

135

weight of a moderately heavy man (800 N). To support the same weight in steel a wire of area 0·4 mm² (diameter 0·71 mm) would be sufficient, but with glass or perspex a rod of area 10 mm² (diameter 3·6 mm) would be required.

3. *Theoretical strength of a solid*

In any practical endeavour it is usually helpful to know something about the theoretical limits of what is possible. Knowledge of this kind provides a basis for estimating the possibilities for advance in any particular direction, and may avoid the wasting of time on fruitless experiments. It is not surprising, therefore, to find that considerable attention has been given to the question of the theoretical strength of materials, and that such theoretical studies have had an immensely stimulating effect on the practical development of high-strength materials.

In calculating the theoretical strength it is simplest to start by considering a single crystal. A classic case is that of rocksalt (NaCl) which crystallizes in a rather simple cubic form. In the crystal lattice neighbouring sites are occupied alternately by positive sodium ions (that is, sodium atoms which have lost one electron) and negative chloride ions (chlorine atoms which have gained one electron). In any one plane of the crystal each sodium ion has four chloride ions as nearest neighbours, and *vice versa* (fig. 9.2). In the complete 3-dimensional array, each Na⁺ ion is surrounded by six Cl⁻ ions.

To calculate the tensile strength we imagine the crystal to be divided into two pieces by the separation of the right-hand side from the left along the plane AB (fig. 9.2). In carrying out this separation a force has

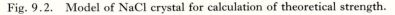

Fig. 9.2. Model of NaCl crystal for calculation of theoretical strength.

136

to be exerted to overcome the attractive forces between each of the ions $a_1, a_2, a_3...$ on the right of this plane and the oppositely charged partners $b_1, b_2, b_3...$ on the left. Hence if we know the attractive force between a single positive and a single negative ion, and also the number of ions per unit area of the surface, we can readily derive the total force required to separate the surfaces.

All the information required for this calculation is available from various sources. The attractive force between equal positive and negative charges, is, according to Coulomb's law, proportional to the square of the charge and inversely proportional to the square of the distance between them. Hence, if x is the distance between the centres of Na^+ and Cl^- ions, this component of the force is given by A/x^2, where A is a constant. This, however, is not the only force operating, for if it were, the ions would be drawn together until they coincided (i.e. $x = 0$). There is, in addition, a *repulsive* component of the force, arising from the positively charged atomic nuclei. This force varies more rapidly with the separation; it is inversely proportional to the *tenth* power of the distance. The net attractive force f may thus be written in the form

$$f = \frac{A}{x^2} - \frac{B}{x^{10}}, \tag{9.1}$$

where B is another constant.

The form of the function represented by equation (9.1) is shown in fig. 9.3. The point $x = a_0$ corresponds to the normal equilibrium dis-

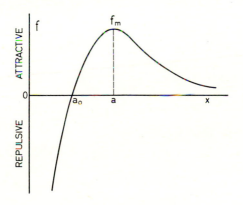

Fig. 9.3. Variation of force between positive and negative ions with distance x between their centres.

tance between the centres of the ions (i.e. the lattice spacing). For $x < a_0$ the force is repulsive, while for $x > a_0$ it is attractive. At a particular separation given by $x = a$, the tensile force reaches its maximum value f_m; this (multiplied by the number of ions per unit area) gives the required tensile strength of the crystal.

Calculations of the theoretical strength have been carried out for a number of other materials, though the precise details of the calculations vary according to the nature of the interatomic force field, etc. The calculation is not limited to the case of a crystal, but may be applied (approximately) to a glassy structure. Values so calculated are compared with actual measured values in the table below.

Material	Theoretical	Observed
Rocksalt (NaCl)	2700	4·4 (bulk crystal)
		1000 (whiskers)
Iron	46000	2000 (steel wire)
		14000 (whiskers)
Glass	8000	80 (sheet)
		1000–3000 (filaments)
Cellulose (crystal)	11000	1000 (fibre)

Table 9.3. Calculated and observed tensile strengths (MN m^{-2}).

In all cases the calculated strength is considerably higher than the strength actually attained. If we exclude ' whiskers ' (very fine single crystals) and glass filaments, which will be referred to later, the difference amounts to a factor of 600 for rocksalt, 100 for glass, 23 for steel and 11 for cellulose. A pictorial representation of this difference, in the case of glass, is given in fig. 9.4. Differences of a similar order have been obtained in all the materials which have been studied.

4. The Griffith theory of flaws

The very large gap between the theoretical and observed values of strength for such a wide range of materials calls for a fundamentally different approach to the question of the theoretical strength. The theory, in the form presented above, is clearly totally inadequate. It must be concluded that a solid does not break by a process of uniform separation of one plane of atoms from another over the whole cross-section of the specimen.

A more realistic basis for the consideration of the problem of strength was provided by Griffith*. Griffith's theory, which was originally published in 1920, was concerned primarily with glass, but his ideas have been applied to materials of all kinds. According to this theory the discrepancy between the theoretical and the observed strength is attributed to the presence of defects or flaws in the material. The simplest case to consider is that of a specimen in the form of a parallel strip containing an edge crack of length c (fig. 9.5), subjected to a tensile stress S. Griffith was able to show that if the stress S exceeds a certain critical value S_k the

* Not the Griffith who put forward the ' skipping-rope ' theory of rubber elasticity (p. 35).

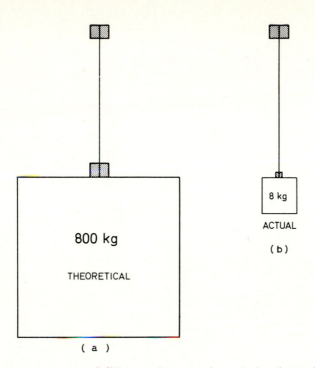

800 kg

THEORETICAL

(a)

8 kg

ACTUAL

(b)

Fig. 9.4.   Representation of difference between theoretical and actual strength of glass.   (a) Theoretical and (b) actual load which may be supported by filament of area 1 mm².

crack will extend (leading to breakdown), but that if it is below this value crack propagation will not occur.   The condition for breakdown involves the energy required to form the new surface, as the crack advances, but this need not concern us at the moment; the important point is that the critical stress diminishes as the crack length increases.   This is expressed by the relation

$$S_k = \frac{\text{const}}{\sqrt{c}}. \qquad (9.2)$$

It follows from this that the breakdown is a catastrophic process.   If the specimen contains an original crack of length c, and the applied stress S is gradually increased, the crack length will remain unchanged until the point is reached at which S just exceeds S_k.   The crack length c will then begin to increase, and this increase will *reduce* S_k, and hence *accelerate* the process of crack growth.

If we accept the Griffith theory, we are in a position to understand why the original calculations of the theoretical strength give such wildly incorrect results.   There is nothing wrong with the actual calculation of the

139

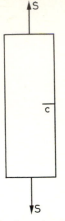

Fig. 9.5.   Application of tensile stress to specimen containing edge crack.

force required the to separate two surfaces; what is wrong is the assumption that the material fractures by simultaneous separation across the whole plane of fracture.   The essential feature in the Griffith theory is the concept of *progressive* fracture by a localized breakdown at the crack tip.   In physical terms, this localized breakdown can be regarded as arising from the actual stress existing in the material at the tip of the crack.   This local stress is very much higher than the average stress in the specimen in regions at a distance from the crack; the presence of the crack therefore has the effect of magnifying or concentrating the stress in the vicinity of the crack tip.   It may be shown mathematically that the amount of this magnification is the greater, the smaller the radius of curvature of the crack tip (other things being equal).   Failure occurs when the true stress at the tip of the crack exceeds the theoretical breaking stress for the material.   Thus the finer the crack, the greater is the weakening effect which it produces.   In the case of a brittle solid the radius of the crack tip is probably of the order of the distance between atomic centres, i.e. about 10^{-7} mm.   With this value of radius stress magnification factors of 1000 or more can easily be obtained.   There is thus no difficulty in accounting for a reduction of strength by a factor of 1000, compared with the calculated strength for an idealized material containing no cracks or flaws of any kind.

The formula (equation 9.2) relating the strength to the depth of the crack was tested experimentally by Griffith, using glass sheets into which cracks of various measured depths had been introduced.   Not only was the inverse square-root relationship found to hold, but the value of the energy required to form the new surface (contained in the constant in equation (9.2)) was found to be about $1 \cdot 0$ J m^{-2}, which is of the order of magnitude to be expected from other considerations.

If the Griffith theory is to account for the actual strength of glass, it is

140

necessary to assume that the original sheet contains 'natural' cracks or flaws of a certain size. This size, which can be estimated by inserting the observed value of strength into equation (9.2), works out at about 0·01 mm or less. There are various lines of evidence to suggest that such cracks, though not directly visible, are in fact present on the surface of ordinary glass. The most direct evidence is obtained by evaporation of sodium vapour on to the surface; a pattern of lines, which can only be cracks, is then seen to cover the surface. Unfortunately it is not certain that these cracks were present in the original glass: they could have been produced by the vapour treatment. The indirect evidence is stronger. Glass freshly drawn, and not brought into contact with any other object, has a strength of 3500 MN m^{-2}—nearly half the theoretical figure—and similar values are obtained by etching or dissolving away the surface in hydrofluoric acid, thus rounding off any sharp cracks.

Toughened glass. By rapid cooling of the surface of a sheet of glass from the molten state by air jets it is possible to produce a compressive stress in the surface layer, which arises from the difference of temperature between the surface and the interior during cooling. This compressive stress will prevent the opening up of cracks at points of potential damage, and it is necessary to apply a tensile stress equal in magnitude to the surface compressive stress before the possibility of crack propagation arises. This 'toughened' glass, which is widely used in car windscreens, has up to three times the tensile strength of ordinary sheet glass. The disintegration which occurs when it does break is the result of the large amount of internal strain energy (associated with the surface compressive stress) which is thereby released.

The effect of diameter. Griffith himself showed that the strength of glass filaments increased as the diameter decreased down to 0·003 mm, and this effect has since been observed by other workers. Clearly the crack size must be less than the filament diameter; this automatically excludes the larger Griffith cracks in the finer filaments. Also, on account of their smaller surface area, the probability of a surface crack of any particular size is correspondingly reduced. It is noteworthy that the strength of the finest filaments is similar to that of etched glass.

'Whiskers', which may be formed in metals and other materials, are a special case of exceedingly fine filaments. They are actually single crystals, having diameters of a few thousandths of a millimetre, and lengths of a few millimetres. They are usually formed by deposition from the vapour in an electrically heated furnace at a very high temperature, and are deposited as a kind of hairy mat. Whiskers have been prepared in a variety of materials, both metallic and non-metallic (e.g. sodium chloride, graphite and the oxides of aluminium and beryllium). They are of special interest not only because their strength approaches the theoretical strength to within a factor of two or three, indicating that they are practically devoid of flaws, but also because of their great practical possibilities as reinforcing or strengthening materials for

141

incorporation into plastics or light metals such as aluminium. Their application, however, is at present very severely restricted by the slow and costly method of manufacture employed.

5. *Glassy polymers*

The brittleness of glass, as has already been noted in Chapter 5, is due to the absence of any mechanism for relieving an applied stress. This is particularly significant in relation to the Griffith theory, owing to the stress-magnifying effect of any cracks or defects in the material. The polymeric glasses are somewhat less brittle than the inorganic glasses because in them some possibility of local molecular movement or re-arrangement under high stresses does exist, whereas in the inorganic glasses it is virtually non-existent. The inorganic glasses are almost perfectly elastic; they are incapable of plastic deformation such as that which occurs, for example, in metals. The polymeric glasses (and polymers generally) are not so perfectly elastic; they show a limited amount of deformation of the plastic or viscous type, which is irreversible and therefore represents a dissipation of elastic strain energy. Such irreversible deformations are a result of the rather weak secondary forces between the chain molecules, which enable a small amount of rotation about single bonds in the chain to take place at temperatures well below the glass-transition temperature.

The presence of a viscous or irreversible component of deformation makes any calculation of the theoretical strength of a glassy polymer on the basis of the separation of one plane of atoms from another even more unrealistic than in the case of an inorganic glass or crystal. It also introduces difficulties in the application of the Griffith theory to the interpretation of the strength of glassy polymers. The Griffith theory is, in principle, applicable only to a perfectly elastic solid. In a glassy polymer the part played by flaws (whether on the surface or within the body of the material) is by no means as clear-cut as in the case of an inorganic glass. Despite these reservations, however, the basic concepts of the Griffith theory have in fact provided most of the stimulation for the study of the strength and fracture properties of the glassy polymers, even though the interpretation of the effects observed may necessitate some modification of these basic concepts.

6. *Fracture surface energy*

In the Griffith theory the strength is related to the crack length c and to the work or energy required to create a new surface (contained in the constant in equation (9.2)). In this section we shall have to consider this surface energy, usually represented by the Greek letter γ (gamma), in more detail. We will therefore write equation (9.2) in the more complete form

$$S_k = \sqrt{\left(\frac{2E\gamma}{\pi c}\right)}. \tag{9.3}$$

This gives the strength S_k in terms of the product $E\gamma$, where E is Young's modulus.

From measurements of the dependence of the strength S_k on the crack length c, this equation enables us to determine $E\gamma$, and hence, since E can be directly measured, to derive the value of γ. Experiments on these lines have been carried out on both perspex and polystyrene. A typical result for perspex is reproduced in fig. 9.6. Despite the reservations noted above, it is seen that the Griffith relationship (equation (9.3)) between strength and crack length applies, within the accuracy of the experimental data. The value of γ derived from these particular experiments was 220 Jm^{-2}. The corresponding figure for polystyrene was 170 Jm^{-2}. These values are of a higher order of magnitude than the values obtained for inorganic glasses (1·0 Jm^{-2}).

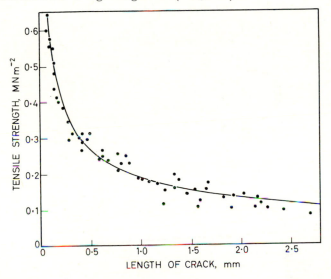

Fig. 9.6. Relation between tensile strength and crack length for specimens of perspex containing edge cracks of different lengths (Berry, 1961). The curve corresponds to equation (9.2).

In the discussion of the theoretical strength of a brittle solid we considered the force required to separate two planes of atoms on either side of the fracture surface. The same model may be used to calculate the total *work* required to separate the surfaces completely, the work being simply the product of the applied force and the distance through which it is moved. The surface work of glass, calculated in this way, is about 1·0 J m^{-2}, in agreement with Griffith's experimentally derived figure. The corresponding experimental figure for perspex, being more than one hundred times as great, cannot possibly be accounted for in a similar manner. The most natural explanation is that the process of

143

fracture is preceded by an extensive amount of viscous or plastic deformation in the vicinity of the crack tip which involves a large dissipation of energy. Direct evidence of such local deformation is provided by the examination of the fracture surfaces in the electron microscope.

In the case of glass, Griffith was able to account satisfactorily for the actual strength by postulating the presence of minute cracks, or natural flaws, of length less than 0.01 mm. Great difficulties arise in the attempt to explain the strength of polymeric glasses in this way. Insertion into equation (9.3) of the experimentally derived value of $E\gamma$ for the case of polystyrene, namely 3.4×10^{12} J^2 m^{-5}, together with the value of the strength for the undamaged material (50 MN m^{-2}) leads to the result that the ' natural ' or inherent crack length is about 1 mm. This conclusion is quite unacceptable. Cracks of such dimensions are certainly not present in the unstressed material, for, if they were, they would be readily seen in the microscope (or even by the naked eye). Whatever the reason for this paradoxical result (for which some possible explanations have been put forward) it certainly indicates that the strength of glassy polymers cannot be interpreted in terms of Griffith cracks as normally understood.

7. *Effect of orientation*

In Chapter 2 reference was made to the ease with which rubber, when stretched and crystallized, may be split along the direction of the orientation. This *fibrosity* is a well-known property of oriented crystalline polymers.

However, crystallization is not an essential requirement for the development of different mechanical properties in different directions. The important factor is the degree of molecular alignment. Crystallization assists in stabilizing the aligned state and therefore tends to accentuate the inherent effects of orientation.

In the glassy polymers a moderately high degree of molecular alignment may be produced by applying a high tensile force to a specimen at a temperature above the glass-transition temperature (i.e. when the material is more or less rubberlike) and then lowering the temperature while the specimen is still under tension. Specimens produced in this way show a strong tendency to split in the direction parallel to the orientation. This tendency to longitudinal splitting arises from a combination of increased strength in the direction of the orientation and reduced strength in the transverse direction. This *anisotropy* of strength, as it is called, is so marked that a specimen of drawn perspex, when subjected to a tensile test *along* the direction of drawing, may actually fail not by propagation of a crack across the specimen (i.e. in a direction transverse to the orientation), but by longitudinal splitting, in a fibrous manner (fig. 9.7). This phenomenon is analogous to the longitudinal splitting of stretched crystalline rubber under impact (Plate 2).

144

Fig. 9.7. Longitudinal splitting of strip of perspex containing molecular orientation, when subjected to tensile stress along the direction of orientation. The specimen is shown bent to display the fibrillar type of fracture (Curtis and Treloar, 1968).

The mechanical anisotropy of glassy polymers arising from molecular orientation is of considerable practical importance in many problems of design. In the moulding of articles in glassy polymers the highly viscous molten polymer is forced under high pressure into the mould, which is then rapidly cooled. In the process of flow the molecules undergo a considerable degree of orientation, which is readily shown up by observations with polarized light. This orientation is then 'frozen-in' on cooling, and is a potential source of weakness if the article is to be subjected to stresses in a direction transverse to the orientation.

8. *Fibres and fibre reinforcement*

As is indicated in the tables on p. 135, the fibres are not only the strongest form of polymers, but compare favourably in strength (calculated on a mass basis) with the strongest metals. This high strength, as we have seen, is associated with a high degree of molecular orientation. The polymer chain itself, which is usually formed of C–C bonds, is inherently very strong, but the strength in the transverse direction, which depends on the weak secondary forces between chains, is comparatively low. By aligning the chains in the direction of the axis of the fibre we make the best possible use of the inherent strength of the molecular chains. Even in the most highly oriented fibres, however, there must remain a fraction of the molecules in the non-crystalline regions which are still imperfectly oriented, or perhaps not oriented at all, and it is probably in these less highly oriented regions that the fracture

145

process starts. This, rather than surface cracks or defects in the ordinary sense, probably accounts for the failure of actual fibres to approach the theoretical strength calculated for the chain direction in the crystal.

The greatest disadvantage of the glassy polymers, on the other hand, is their inherent weakness and brittleness. It is therefore natural to consider the possibility of incorporating fibres into glassy polymers and thus producing polymers in bulk form of greatly increased strength. This increase in strength may be expected to be proportional to the fraction of the total volume occupied by the fibres; for volume fractions exceeding 50% (which are quite practicable) the strength of the composite material should therefore begin to approach the strength of the fibre itself.

For a fibre to have a strengthening or reinforcing effect, it is necessary for it to resist the applied stress more effectively than the material in which it is embedded. This requirement implies that *both* the modulus *and* the strength of the reinforcing fibre shall be higher than the corresponding modulus and strength of the polymer which it reinforces. On both these counts the inorganic fibres, notably glass, asbestos and carbon, are the most generally suitable. By far the largest proportion of commercially produced reinforced polymers utilizes glass fibres. The matrix in which the glass fibres are embedded may be one of a number of polymers, the commonest being certain types of polyester resins containing double bonds and the so-called *epoxy* resins. The glass fibres may be arranged randomly as two-dimensional sheets or mats, or they may be incorporated as sheets of woven fabric. Several layers of sheet may be superposed to form a multiple glass-fibre *laminate*. The reinforced sheet can be moulded or pressed into any desired form before the final setting process.

The properties of the composite structure are determined mainly by the properties of the reinforcing fibre. Some improvement may be expected by the use of glasses of higher modulus and higher strength than ordinary glass, but attention is mainly being directed to the development of new materials, for example, boron fibres (deposited on a tungsten core), carbon filaments, and 'whiskers' of various materials.

Carbon filaments, whose strength exceeds that of steel (table 9·2), are of particular interest. The carbon in these filaments is in the form of very small crystals, having a structure closely resembling that of graphite. Ordinary graphite is a very soft material, often used as a lubricant because of its ready breakdown under stress. The graphite crystal is composed of layers or sheets of hexagonally packed and chemically-bonded carbon atoms stacked one above the other (fig. 9.8). The strength within the sheet is very high but the forces between the sheets are very weak. Consequently, the crystal has a very high strength in the plane of the sheet, but a very low resistance to the sliding or shearing of one sheet over another. In ordinary graphite the crystals have no particular orientation, hence sliding will occur in those crystals which are

146

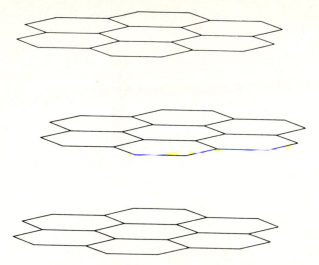

Fig. 9.8.   Crystal structure of graphite.

appropriately oriented with respect to an applied stress.   In carbon filaments as used for reinforcement, on the other hand, the strong hexagonally packed sheets within the crystalline parts of the fibre are all oriented in one direction, the direction of the fibre axis, thus conferring on the fibre the high inherent strength of the crystal in the planes of the sheets.   The way this is achieved is remarkable and rather unexpected. The filaments are actually produced from a highly oriented polymer fibre.   The fibre used is one of the acrylic fibres.   This is decomposed first of all at a comparatively low temperature under conditions such that ring formation occurs by chemical reactions between adjacent chain molecules, without loss of axial orientation.   This is followed by treatment at a very much higher temperature to produce the final graphite structure.   This remarkable process was developed at the Royal Aircraft Establishment, Farnborough, where, incidentally, the original work of Griffith was carried out.

9. *Conclusion*

Summing up the main ideas presented in this chapter, we may note again that while strength is a property of the greatest importance in the practical use of materials, the factors which actually determine strength are complicated, and in some respects rather more elusive than those which determine most of the other properties of interest to the designer. Any theory relating to this subject is at present necessarily somewhat tentative, in the nature of a working hypothesis rather than of an established certainty.   This means that the subject of strength and fracture is still at a comparatively early stage of development.   Nevertheless, it is

147

hoped that the reader will have gained sufficient insight into the kind of problems involved, and into the nature of present scientific developments, to have acquired some feeling for the subject, and some appreciation of its potentiality for development.

The great upsurge in the polymer industry which the present generation has witnessed has taken place largely within and around the *chemical* industry of this country and of the world at large. It has been concerned largely with the *production* of different types of polymeric materials, regarded as chemical species. The next generation is likely to see a shift of emphasis away from the chemical processes of production, which to an increasing extent are being brought under control, towards the *uses* of polymers as structural and engineering materials. It is in this area that our knowledge is still inadequate, and indeed in many respects rudimentary. It is in this area, primarily, that we may look forward to the greatest concentration of future scientific effort. Within this general area, a large fraction of the total effort will undoubtedly be devoted to the problems of strength and fracture.

CHAPTER 10
water absorption and swelling

1. The process of water absorption

THE capacity for swelling by the absorption of water is shared by an enormous variety of polymeric substances. One important group includes the proteins, whether in the form of fibres (wool and hair), in bulk (gelatin or glue), or as a component of living matter (peas, beans and other seeds). Another includes the various forms of cellulose: wood, cotton, flax, and regenerated cellulose or rayon. The processes of life are carried on in a water-absorbing environment, and the control of the water content of the tissues is a fundamental element in the growth and development of every living being, from the unicellular organism or amoeba to man.

The absorption of liquids by polymers is what is called an *osmotic* process. This is closely related to the process of solution. It is, in fact, a rather special form of solution, and, as with other forms of solution process, it involves the diffusion of one species of molecule into a medium composed of a different species. We usually think of solution in terms of the diffusion of a solid (e.g. sugar) into a liquid (water), to form a homogeneous molecular mixture. In the swelling of a polymer it is the liquid which diffuses into the solid to form a molecular mixture which is in all essential respects a solution. In the more familiar type of osmosis as encountered, for example, in the measurement of osmotic pressure (Chapter 2) the pure liquid is separated from the solution by a semi-permeable membrane which permits the diffusion of the liquid molecules into the solution but prevents the transfer of the molecules of the solid in the reverse direction. The only difference in the case of the swelling of a polymer is that the mixed phase (polymer plus liquid) happens to be a solid, and, since the polymer molecules cannot be individually detached from the solid structure and so pass into solution, there is no necessity for a semi-permeable membrane; the swollen polymer is automatically semi-permeable. (This is not strictly true for certain systems, e.g. un-vulcanized rubber, whose molecules may diffuse slowly into the solvent; in such systems the swelling which precedes solution depends on the great difference in *rates* between the two processes. The liquid diffuses rapidly into the polymer, but diffusion of the polymer into the liquid is extremely slow.)

The absorption of the liquid molecules by the polymer does not necessarily require that the polymer should be in physical contact with the liquid; it is sufficient that it should be in contact with liquid molecules in the form of vapour.   This is why cotton or woollen fabrics become damp even without contact with water; they absorb moisture from the water vapour which is always present in atmospheric air.   If a mass of cotton fibres (cotton wool) is sealed into a bulb A (fig. 10.1) connected to an-

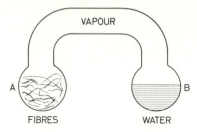

Fig. 10.1.   Absorption of water from vapour phase.

other bulb B containing water, the cotton fibres will absorb exactly the same amount of water as if they were immersed in the water.   There is, however, one difference.   If the mass of fibres is immersed in water it will retain water in the condensed or liquid form on the surface of the fibres, and in the minute channels or ' capillaries ' between fibres, in addition to that molecularly absorbed within the fibres.   In the case of a woven or knitted fabric the amount of water held in this way by capillary attraction or surface tension may exceed the amount actually absorbed by the fibres, but this water is removable by centrifugation (' spin-drying ') and plays no part in the molecular process of swelling.

We are all aware that the moisture uptake of a fabric depends on the amount of water vapour in the atmosphere.   Clothes which have been left in a cold damp room contain a considerable amount of water, even without being superficially ' wet '.   This is because of the high humidity or water-vapour content of the cold air.   On being brought into a warm, dry atmosphere the reverse process occurs; the water diffuses out of the fibres into the surrounding air.

The dampness or humidity of the atmosphere is measured in terms of what is called the *relative humidity*.   This is the ratio of the amount of water in the air to the maximum amount which the air can hold, at the particular temperature considered; it is usually expressed as a percentage. Thus perfectly dry air would correspond to 0% relative humidity (R.H.) and completely saturated air to 100% R.H.   There are several ways of measuring the relative humidity.   The most direct is to extract the whole of the water from a measured quantity of air by means of a power-ful drying agent such as phosphorus pentoxide, and to weigh the amount of water taken up.   Less fundamental but more convenient methods are the wet-and-dry bulb thermometer, and the hair hygrometer.   The former depends for its action on the removal of heat from the wet bulb of

150

the thermometer due to the latent heat of vaporization of the water; the lower the relative humidity, the higher is the rate of evaporation and hence the greater the drop in temperature. The hair hygrometer depends on the relation between the water content of hair and the relative humidity (discussed below); the absorption of water increases the length of the hair, and this increase in length is used to actuate a lever which records the relative humidity on a suitably calibrated scale.

The reader is probably familiar with the fact that table salt becomes damp on exposure to a damp atmosphere. The vapour pressure of a concentrated salt solution is only 75% of that of pure water. Hence (since the water content is proportional to vapour pressure) the equilibrium water content of air in contact with concentrated salt solution (e.g., in a closed bottle) will correspond to 75% relative humidity. If the air humidity exceeds 75%, water will therefore condense on to the salt, but if the R.H. is less than 75% the salt will remain dry.

This property of a salt may be adapted to the determination of the amount of water absorbed by a fibre at any given humidity. If, for example, the water in the bulb B (fig. 10.1) is replaced by a saturated solution of salt (NaCl) the fibres in A (given sufficient time) will eventually absorb a quite definite amount of water, corresponding to the relative humidity of 75%. This equilibrium water content may be measured by direct weighing, preferably without removal of the material from the enclosure. (This may be done by suspending the material from a fine coil spring of silica glass.) Different salts give different saturation vapour pressures, hence by choosing a range of different salt solutions it is possible to explore the whole range of humidity.

The resulting relation between water content and relative humidity for two forms of cellulose, cotton and viscose rayon, on going through a cycle of humidity from complete dryness (0%) up to saturation (100%) and back, is shown in fig. 10.2. Two features of these curves are of interest. First, the relation between water content and R.H. is markedly non-linear: the water content is not simply proportional to the R.H. Secondly, the curves obtained on the second half of the cycle, i.e. with *decreasing* humidity, lie *above* those for the half-cycle of increasing humidity. The water absorbed is less easily removed than it would be if the process were perfectly reversible. This effect is known as *hysteresis*.

2. *Swelling pressure*

It is a remarkable fact that if a polymer is in contact with a swelling liquid, the diffusion of liquid into the polymer will continue to take place even against a considerable hydrostatic pressure tending to oppose its entry. If the polymer is packed into a closed vessel which is permeable to the liquid (e.g. an earthenware pot), the absorption of liquid will generate a pressure on the walls of the vessel which may reach a very high value. This 'swelling pressure' has very important practical consequences. Timber incorporated into a building in a dry state will sub-

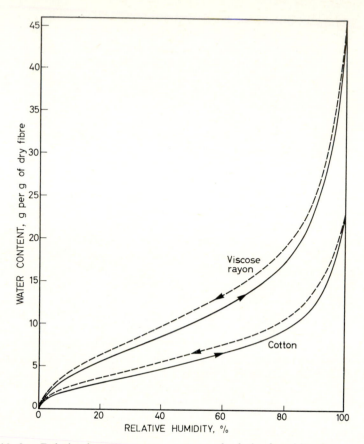

Fig. 10.2. Relation between water content and relative humidity for cotton and viscose rayon. The direction of change of humidity is indicated by the arrows (Jeffries, 1960).

sequently swell on exposure to atmospheric dampness; the resultant stresses may cause jamming of doors, window frames, etc., or other troubles. Corn and other grains (which contain starch, a polymer closely allied to cellulose, as well as proteins) behave in a similar manner, and present problems in bulk storage or transport, particularly in shipping. Entry of water into a cargo of grain leads to a build-up of pressure which may be sufficient to burst open the holds, thus endangering the stability of the ship.

The amount of liquid absorbed by a polymer depends on the applied pressure in a reversible manner. If the swelling is restrained by the walls of the retaining vessel, the pressure will rise to a particular value, which depends on the liquid content of the swollen polymer. This pressure is termed the ' swelling pressure '; it is the pressure which has

152

to be applied to the swollen polymer (not to the liquid) to prevent further entry of liquid. If a pressure greater than this is applied, liquid will be ' squeezed out ' of the polymer. The mathematical expression for the swelling pressure P, for any given degree of swelling, involves the relative vapour pressure (or relative humidity in the case of water) which would be in equilibrium with the swollen polymer (at the same degree of swelling) in the absence of an applied hydrostatic pressure. The relative vapour pressure is defined as p/p_0, where p is the pressure of the vapour in equilibrium with the swollen polymer, and p_0 the saturation vapour pressure of the pure liquid. This relation is

$$P = -\frac{RT}{V_1} \log_e \left(\frac{p}{p_0}\right), \qquad (10.1)$$

where R is the gas constant per mole, T the absolute temperature and V_1 the volume occupied by one mole of the swelling liquid. In the case of water the relative humidity is substituted for p/p_0.

The following table gives values of swelling pressure for various values of R.H., calculated from this equation, together with values of water absorption (at the corresponding R.H.) for the particular case of viscose rayon (cf. fig. 10.2). Thus, for example, at 50% R.H., this material will absorb 10% of water, in the absence of external pressure. If placed in contact with liquid water (or vapour at 100% R.H.) the pressure required to prevent further swelling is 948 atmospheres. This is a very high pressure, comparable to that at the bottom of the deepest oceans. As the liquid content is reduced, the swelling pressure rises more and more rapidly.

Relative humidity %	P atmos	Water content, % (viscose rayon)
0	Infinite	0
1·0	6300	0·5
10	3150	4
50	948	10
80	301	18
90	144	25
95	70	30
99	14	34
100	0	35

Values of swelling pressure P in atmospheres for various values of relative humidity. Column 3 gives the corresponding values of water content for viscose rayon

3. Heat of wetting

The absorption of water by cellulose is accompanied by a strong evolution of heat. This heat may be measured by immersing a mass of fibres

153

(e.g. cotton wool) in water in a sensitive calorimeter. The *heat of wetting* is defined as the amount of heat given out, calculated on unit mass of dry fibre, on completion of the process of water absorption. Naturally this depends on the amount of water already present in the fibre; the greater this is, the smaller is the further amount of water absorbed, and the smaller, therefore, the heat evolved. The experimental relation, for viscose rayon, is shown in fig. 10.3. The total heat evolved on complete wetting of the dry cellulose amounts to as much as 107 joule (25·5 calories) per gram of fibre. Similar results are obtained for the wetting of wool or hair.

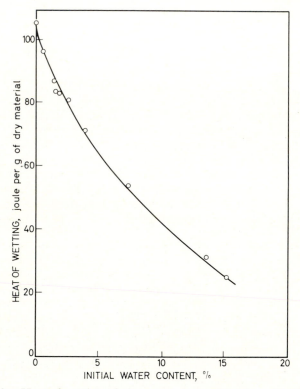

Fig. 10.3.   Heat of wetting of viscose rayon as a function of initial water
content (Guthrie, 1949).

The heat of wetting is defined as the heat evolved on absorption of *liquid* water. If the water is absorbed by contact with water vapour in the atmosphere rather than by direct wetting, there is an additional evolution of heat equal to the latent heat of condensation of water; this amounts to a further 2445 joule (585 calories) per gram of water absorbed. The resultant total heat evolution will depend on the amount of water which the fibre can absorb; in the case of viscose rayon or wool, which

154

absorb about 35% of water, the total heat evolved amounts to about 965 joule (230 calories) per gram of dry fibre.

These thermal effects are entirely reversible, and on drying the fibre an equivalent amount of heat has to be supplied.

4. *Application to clothing*

The water-absorbing properties of fibres are very important in relation to the suitability of particular fabrics for clothing purposes. Materials differ enormously in their capacity to absorb water. The natural fibres, cotton and wool, and also viscose rayon, are very hygroscopic, but the synthetic fibres generally absorb only relatively small amounts of water; typical examples are shown in fig. 10.4. The capacity for absorbing water has certain advantages as well as disadvantages. Among the advantages is the ability to absorb a certain amount of the body's perspiration without becoming saturated. This property, however, is intimately bound up with the *transpiration* of water vapour through the fabric, which depends more on the openness of the yarn and fabric structure, i.e. on the diffusion of moist air through the material, than on the properties of the fibre material itself. On the other hand, hygroscopic materials require longer drying times, which may be a disadvantage in laundering. Also, the evaporation of water from damp cotton or woollen clothing can cause a serious loss of heat from the body.

Another factor of some importance is the effect of absorbed water on the electrical conductivity of the fibre; the hygroscopic fibres are normally good conductors, but the non-absorbent materials are poor in this respect. This may give rise to frictional or ' static ' electrification. The yarn or fabric may become electrically charged by friction during processing or in use, and in materials of low conductivity these ' static ' charges may remain on the surface for quite a long time. This may lead to difficulties in spinning, weaving, etc., arising from the repulsive forces between charged filaments or yarns. In certain types of use—e.g. in the presence of inflammable vapours, such static electrification can be a source of danger, for it may lead to discharge by sparking which is capable of causing an explosion. (This is important, for example, in hospitals, where sparks from clothing may ignite an anaesthetic such as chloroform.) A less dramatic but also quite serious consequence of static electrification is excessive soiling of clothing. The charged surface attracts particles of dust in the atmosphere to itself. These dust particles become firmly attached to the fibre, and are never completely removed by washing. The result is a gradual discoloration of the material. Many of these unpleasant effects can, however, be considerably alleviated by suitable treatment, chiefly by the application of ' anti-static ' agents which increase the surface conductivity of the fibres and so prevent the accumulation of electric charges.

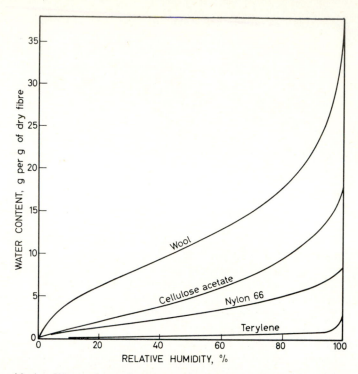

Fig. 10.4.   Water-absorption curves for wool, cellulose acetate, nylon 66
and Terylene (Jeffries, 1960).

5. *What determines the water absorption?*

We have not yet considered the question of the molecular factors which
determine the capacity of a polymer for absorbing water.   This is closely
related to the question of solubility.   Very broadly, molecules of similar
chemical composition tend to attract one another, and hence to mix freely
or form solutions one with another, while molecules of very different
chemical composition tend to avoid each other's company, and hence to
separate out into distinct phases.   A good example is provided by
benzene (C_6H_6) and the low-molecular paraffins (C_nH_{2n+2}).   These are
chemically very similar, and paraffin (or petrol) and benzene are mutually
soluble in all proportions.   On the other hand water, H–O–H, is chemically very different, and if water is shaken up with petrol or benzene, it
will not mix molecularly, but will settle out at the bottom of the vessel as
a separate phase.   The sugars, on the other hand, are chemically rather
similar to water; they contain the atoms of hydrogen and oxygen in the
same proportions as in water, H_2O.   (They are called *carbohydrates* for
this reason.)   The sugars are, of course, extremely soluble in water.
For the same reason cellulose ($C_6H_{10}O_5$), which is also a carbohydrate and

156

contains three OH-groups per glucose unit (fig. 8.9), also has a natural affinity for water, and it is only because of its polymeric nature and high degree of crystallinity that it does not dissolve. As we have already seen, the attraction between cellulose and water may take the form of ' hydrogen bonding ' between the water and the OH-groups in the cellulose molecule (cf. p. 126). The large amount of heat evolved on wetting is further evidence of this rather strong chemical attraction.

The amount of water absorbed depends not only on the chemical nature of the polymer but also on its physical structure. In the absorption of water by cellulose, X-ray and other studies have shown that the structure of the crystallites remains unchanged. The water is absorbed only by the non-crystalline or amorphous component of the structure. This is the reason for the higher absorption of viscose rayon, compared with the chemically identical cotton. In Chapter 7 we saw that cotton was much more highly crystalline than rayon, the amorphous component being estimated at 60% for rayon, but only 30% for cotton. The water absorptions of these two materials are in approximately this same ratio, namely 2 : 1 (fig. 10.2).

The hydrocarbon fibres, polyethylene and polypropylene, are completely incompatible with water. This is in harmony with the incompatibility of water with paraffinic materials generally, which we have already noted. Nylon is intermediate in properties; it contains the CO–NH group present in wool and other proteins, which is strongly attractive to water, but this is interspersed with a succession of CH_2 groups, which are inherently water-repellent. The result is that nylon is only slightly water-absorbent (fig. 10.4). Terylene has a more strongly hydrocarbon nature than nylon, and hence absorbs even less water.

6. *The swelling of rubbers*

The solubility of rubbers, particularly the common hydrocarbon rubbers such as natural rubber, butadiene-styrene rubber and butyl rubber, in ordinary solvents such as petrol and benzene, is a further illustration of the principle referred to in the preceding section. If, however, the rubber is cross-linked all the molecules are joined together chemically in the form of a network. This network cannot disperse, it can only swell. A vulcanized rubber is therefore insoluble.

We have already seen that the process of swelling involves the diffusion of the liquid molecules into the polymer. This diffusion is made possible by the freedom of motion of local segments of the polymer chains, which enables the polymer to accommodate the incoming liquid molecules. For small amounts of absorbed liquid the process is not significantly different from that involved in the diffusion of one liquid into another, e.g. benzene into paraffin. But as the amount of liquid diffusing into the rubber increases, the effect of the polymer structure comes into play, the swelling leading to an expansion of the network

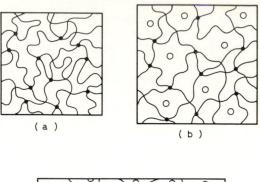

(a)

(b)

(c)

Fig. 10.5.   Expansion of cross-linked network on swelling.   (a) Unswollen.
(b) Moderate degree of swelling.   (c) High degree of swelling.

structure (fig. 10.5).   As a result of this expansion the polymer chains are
extended in much the same way as if they were subjected to an external
force, except that in the case of swelling the expansion is in three dimensions (i.e. a volume expansion) rather than in one dimension.   The effect,
however, is similar; the expansion is resisted by the tendency of the
chains to contract to a shorter length.   The balance of these two effects,
(1) the tendency of the molecules of liquid to mix with or diffuse into the
polymer molecules and (2) the tendency of the network to contract to a
smaller volume, leads to an equilibrium state corresponding to a precisely defined liquid content or equilibrium degree of swelling.

The final or equilibrium degree of swelling depends on the extent to
which the already existing cross-linked network can be expanded.   In
considering the elasticity of a cross-linked rubber (Chapter 4), we saw
that the deformability under an external stress, i.e. the value of the elastic
modulus, depends on the degree of cross-linking; the higher the degree of
cross-linking, the higher is the modulus, and the smaller the amount of

158

deformation under a given stress. The same argument applies to the extension of the network by internal swelling. The higher the degree of cross-linking, the greater is the resistance to network expansion, and the smaller the degree of swelling. The quantitative theory of this effect was worked out independently by Flory and by Huggins in 1942, and is known as the Flory–Huggins theory. This theory leads to a number of interesting conclusions, all of which have been confirmed experimentally in all essential respects. It predicts, for example, a definite relation between the degree of swelling in a particular solvent and the modulus of elasticity of the unswollen rubber. For high degrees of swelling this relation takes the form

$$\frac{GV_1}{RT} = \frac{A}{Q^{5/3}}, \tag{10.2}$$

where G is the shear modulus of the unswollen rubber, Q the volume swelling ratio (i.e. the ratio of the volume in the swollen state to the volume of the dry rubber) at equilibrium swelling, V_1 is the molar volume (volume of 1 gram-molecule) of the solvent, and A a constant whose value is dependent on the particular rubber-liquid system studied. Taking logarithms, equation (10.2) may be expressed in the form

$$\log G = \log \frac{A}{V_1 RT} - \frac{5}{3} \log Q \tag{10.3}$$

or

$$\log G = \text{const.} - \frac{5}{3} \log Q, \tag{10.4}$$

since the temperature T is assumed constant. Thus if $\log G$ is plotted against $\log Q$, a straight line of slope $-\frac{5}{3}$ should be obtained. Figure 10.6 shows the result of an experiment carried out by Flory on a series of variously cross-linked butyl rubbers swollen in cyclohexane. The line drawn through the experimental points has the slope $-\frac{5}{3}$.

A rather more surprising conclusion derived from this theory is that if the rubber is swollen while under a tensile stress, the equilibrium degree of swelling should be *increased*. This theoretical result is of particular interest, because an effect of this kind had not previously been looked for, and indeed, it is contrary to general expectations. Figure 10.7 shows the result of an experiment carried out by the author on natural rubber swollen in heptane (C_7H_{16}). The point A represents the stress-free swollen state; application of an extension caused the swelling to increase. The experimental points fall very closely about the calculated curve. The effect of the extension on the swelling is completely reversible; on removal of the stress the degree of swelling returns to its original value.

The effect of tensile strain on the swelling equilibrium may be directly demonstrated by allowing a strip of rubber (say 3 mm thick) to swell to equilibrium in benzene (which will take several hours) under a strain of 100% or more (referred to the unswollen dimensions). If the sample is

159

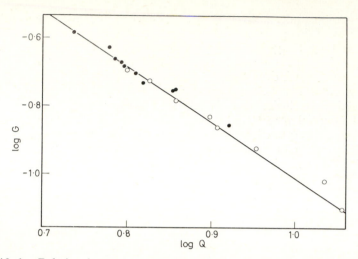

Fig. 10.6. Relation between equilibrium degree of swelling (Q) and elastic modulus (G) for a series of differently cross-linked butyl rubbers swollen in cyclohexane (Flory, 1946). (G in MN m^{-2}.)

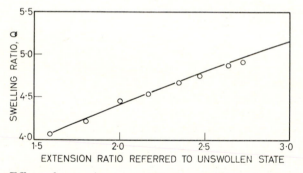

Fig. 10.7. Effect of extension on equilibrium swelling of vulcanized rubber. Points experimental; curve theoretical.

then removed from the liquid bath and quickly dried with blotting paper, release of the stress will lead to a visible re-wetting as the benzene exudes from the rubber.

The effect of a tensile stress on the equilibrium degree of swelling can be understood in a general way if we consider the problem in relation to the swelling pressure. In the latter case the applied stress is a hydrostatic pressure. For any given value of the pressure there is a corresponding equilibrium value of the swelling. If the pressure is reduced below this value a further quantity of liquid is absorbed; if it is increased, liquid is forced out of the swollen polymer. A rather similar result would be obtained if, instead of applying an ' all-round ' or hydrostatic pressure, we applied a pressure to the polymer in one direction only, that is, by squeezing it between parallel plates. Again, increasing the pressure would reduce the amount of swelling and *vice versa*. But an extension in one direction is the exact opposite of a compression, hence it is not surprising that a tensile stress has the opposite effect on swelling, leading to an *increase* of swelling with increasing stress.

7. *Contrast with water-absorbing polymers*

While the basic mechanism of swelling is the same, there are important differences between the phenomena of swelling in rubbers and the corresponding effects in water-absorbing substances. The first and most important characteristic difference is that in the case of rubbers there is no specific *attractive* force between the molecules of the rubber and the swelling liquid. The polymer molecule is, as it were, *neutral*; it has no particular preference for the company of a liquid molecule or for that of a segment of another polymer molecule—the two are so similar that they are hardly distinguishable. The benzene molecules diffuse into the rubber in much the same way as they would diffuse or disperse in liquid paraffin; there is no resistance to their entry, but neither is there any particular attraction. This is in marked contrast to the behaviour of water with respect to cellulose; in this case, as we have seen, there is a strong attractive force amounting almost to a chemical binding. Closely related to this difference is the absence of any significant evolution of heat on swelling. In the case of rubber and benzene, for example, there is actually a slight *reduction* of temperature on swelling.

The second important difference is that rubbers swell to a much greater extent than most water-absorbing polymers, particularly fibres. This is because of the amorphous structure of the rubbery polymers. In crystalline polymers the absorption of liquid is effectively limited to the non-crystalline component, and the swelling of this component is restrained by the interlocked system of crystallites, which exerts an effect similar to that of a very highly cross-linked network. In fibres the maximum swelling corresponds to a water uptake of about 35%, but in rubbers there is no difficulty in obtaining swellings of up to 10 times the original volume.

8. *Oil-resistant rubbers*

Owing to the general nature of the swelling phenomenon, the problem of producing rubbers which are resistant to liquid hydrocarbons—particularly oil and petrol—is a rather difficult one. The problem is important because of the use of rubber components in a variety of engineering applications, for example, for anti-vibration mountings, for valves in pumping systems, and in various sealing devices such as are used in aircraft hydraulic systems, as well as for tanks and flexible containers of various kinds. Resistance to the absorption of a particular liquid can only be obtained by the incorporation into the rubber molecule of a particular chemical grouping which does not readily accept or mix with the liquid. Thus polychloroprene (neoprene), in which the CH_3 side-group of natural rubber is replaced by a chlorine atom, giving the repeating unit

$$-CH_2-CH=\overset{\overset{\textstyle Cl}{\textstyle |}}{C}-CH_2- \qquad (10.5)$$

swells only to the extent of 20 to 50% in mineral oils. Still more effective is the incorporation of the CN-group, as in nitrile rubbers; these contain the unit

$$-CH_2-\underset{\underset{\textstyle CN}{\textstyle |}}{CH}- \qquad (10.6)$$

and will absorb only quite small quantities of oils (1 to 10%). However, the incorporation of the CN-group also has the effect of raising the glass-transition temperature (cf. Chapter 5). The nitrile rubbers therefore suffer from the disadvantage that the temperature range over which their rubbery properties are retained is rather restricted.

9. Conclusion

In the early days of colloid science the phenomenon of swelling appeared as a rather curious property of certain special materials. The understanding of the phenomenon has come about more or less simultaneously with the formulation of the concept of a high polymer. By the application of thermodynamics it became possible to see the essential similarity between the process of swelling in polymers and the more familiar processes of solution as exhibited by low-molecular materials. Even so, the study of water-absorbing materials, which grew up earlier than the study of swelling in rubbers, has tended to be rather self-contained, and to emphasize the factor of chemical attraction or bonding almost to the exclusion of a consideration of the whole mechanism of diffusion and swelling. It is only since the development of the statistical theory of rubber elasticity, and the Flory–Huggins adaptation of this theory to the problems of swelling and solution in rubbery polymers, that the present broader understanding of the relation between the phenomena of swelling and the molecular structure of the polymer has come about.

elastic liquids*

THE usually accepted classification of materials into solids, liquids and gases, though convenient for most purposes, fails to do justice to the wide variety of properties exhibited by actual materials. This is particularly true in the case of solids and liquids, the distinction between which is by no means as clear-cut or as meaningful as the above classification suggests. As an example, let us consider the properties of a very common substance, bitumen (or pitch). On a short time-scale this material behaves like a solid; it has a definite shape, and can be deformed elastically. If struck with a hammer it shows a characteristic brittle type of fracture resembling that of glass. But if left for several days, or weeks, it begins to flow, and in due course it will spread out to form a smooth horizontal surface. In this respect it has the properties of a liquid, differing from an 'ordinary' liquid only in having a very much higher viscosity. By any definition of a liquid it must be included in this category also. It follows, therefore, that bitumen cannot be classified unambiguously as either a solid or a liquid, but has properties common to both these states of matter.

Closer examination of the properties of materials reveals that most 'solids', in fact, show some degree of fluidity, or irreversible deformation under stress; provided this is sufficiently small, however, (as in steel or concrete) this fluidity may be disregarded for purposes of classification. It is only when the effects of solidity and fluidity are of comparable magnitude that the usual classification becomes demonstrably inadequate.

Just as most (perhaps all) solids show some degree of fluidity, so, conversely, many liquids possess some degree of 'solidity' or rigidity. Solidity implies the tendency to retain a fixed shape, or to return to a particular unstrained shape after the removal of an applied stress. In general, it is only in liquids of rather high viscosity that a significant degree of rigidity or elasticity is apparent. However, as in the case of bitumen, the degree of solidity which may be observed is a function of the time-scale of the observations; for short times of observation any inherent elastic properties are accentuated, while for longer times the effects of flow predominate. The reason for this is simply that any deformation of an elastic character is essentially independent of time, whereas the viscous deformation increases in proportion to the time of application of the stress; hence if both elastic and viscous componenis of

---

* The title of this chapter is taken from the book *Elastic Liquids* by A. S. Lodge (Academic Press, 1964).

deformation are present the viscous effects tend increasingly to ' swamp ' the elastic effects as the time-scale of the observations is increased. In many liquids significant elastic effects, which are not apparent under normal conditions of observation, may be revealed by the application of an oscillatory or alternating stress of high frequency; under these conditions the time-scale of the observations corresponds to the period of oscillation, which may be, for example, as short as 10^{-5} s.

In molten polymers and polymer solutions the effects of elasticity which accompany the process of flow are very much more conspicuous and universal than they are in ordinary liquids. There are two reasons for this. The first is their very high viscosity, which, as we have just noted, tends to favour the appearance of solid-like properties and to increase the time-scale within which these properties may be observed. The second is that in polymeric materials the elasticity is *rubberlike* in character, and the magnitude of the elastic effects to be observed is therefore very much greater than in the commoner low molecular-weight materials, which are capable of only very small elastic deformations.

It will be appreciated that the effects of elasticity (solidity) and viscosity (fluidity) cannot be entirely separated, since they are both present simultaneously. The combination leads to some very remarkable effects, of a kind not observable in ordinary liquids, which it is the purpose of this chapter to examine.

2. *Liquids which bounce and liquids which stretch*

A simple example of elastic behaviour in liquids is provided by the silicone polymer known as ' bouncing putty '. The backbone structure of the silicone chain consists of alternate silicon and oxygen atoms. To each of the silicon atoms is attached a pair of hydrocarbon side-groups, which in the case of the commonest of the silicones are simply methyl groups, thus

$$-\overset{\displaystyle CH_3}{\underset{\displaystyle CH_3}{\overset{\displaystyle |}{\underset{\displaystyle |}{Si}}}}-O-\overset{\displaystyle CH_3}{\underset{\displaystyle CH_3}{\overset{\displaystyle |}{\underset{\displaystyle |}{Si}}}}-O- \qquad (11.1)$$

Whereas most polymers in the molten or fluid state are extremely sticky, the silicones are remarkable for their lack of stickiness. (They are used extensively for the treatment of metal or other surfaces to reduce adhesion.) This property makes it possible to demonstrate bouncing properties which, though inherent in other rubberlike polymers, cannot be so readily observed because of the general tendency of such materials to adhere to any surface with which they come into contact.

' Bouncing putty ' is a silicone polymer which, as its name implies, looks and feels rather like putty. It can be moulded in the hands to any form. Unlike putty, however, it does not keep its shape after moulding, and if put into a beaker, it will flow until a perfectly smooth horizontal surface is attained. For long times of observation it behaves like a liquid, and will flow out of a beaker in a continuous smooth stream, like a very viscous treacle (fig. 11.1 (*a*)). It is sufficiently fluid to run out through the gap between an ordinary tin box and its lid. Yet it may be ' gathered together ' and moulded into a ball in the hands, and the ball so formed will bounce like rubber if thrown to the floor (fig. 11.1 (*b*)).

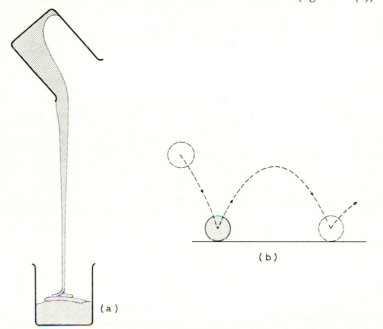

(b)

(a)

Fig. 11.1. (*a*) Gradual flow of bouncing putty over prolonged time of observation. (*b*) Elastic behaviour for short time of observation.

The bouncing properties of this silicone are due to the presence of a network of entangled long-chain molecules which has elastic properties similar in principle to those of rubber in the unvulcanized state. These entanglements, however, have a considerably shorter life-time than in rubber, so that if left to itself for a few minutes the material flows like a liquid.

Elastic properties may be exhibited not only by polymers in the undiluted state but also by polymer solutions, particularly in the range of concentration from 1 to 10%. This elasticity may be revealed by the behaviour of any air bubbles which may be present. If the vessel containing the solution is given a quick rotation the air bubbles may be seen to oscillate when the rotation is stopped. In a more refined experiment designed to study such elastic recoil effects, the solution is contained in a light cylindrical vessel C into which a fixed inner cylinder B is inserted

165

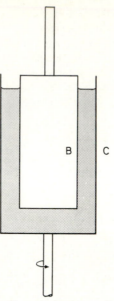

Fig. 11.2.   Demonstration of elastic recoil in polymer solution.

(fig. 11.2).   If the outer cylinder is rotated at a constant rate for a few seconds and then suddenly disconnected from the driving mechanism, it will show an instantaneous rotation in the opposite direction.   This elastic recoil may amount to several degrees.

A rather more striking demonstration of elastic effects is illustrated in fig. 11.3.   In this, a stream of the solution issuing slowly from a bottle is cut with scissors at the point A.   After cutting, the two cut ends of the stream rapidly separate, the top part moving upwards towards the bottle. This effect is well shown by a 4% solution in dekalin of aluminium di-laurate, a compound of aluminium with lauric acid $[CH_3(CH_2)_{10}COOH]$.

Still more curious is the self-siphoning behaviour of a solution of poly (ethylene oxide) $(-CH_2-CH_2-O-)_n$ in water, illustrated in fig. 11.4. The solution is originally contained in the beaker A.   The process is started by pouring a little of the solution into another beaker C which is then carefully lowered to the floor, while A is placed on the edge of the bench.   The solution will then continue to flow at a fast rate from A to C until A is empty.   This self-siphoning property is not fully understood, but it appears to be closely related to the high elasticity of the polymer solution.

3. *Viscous flow in polymers*
General considerations
We turn now to the consideration of the purely viscous aspect of flow in polymers.   This is of fundamental importance in almost every aspect

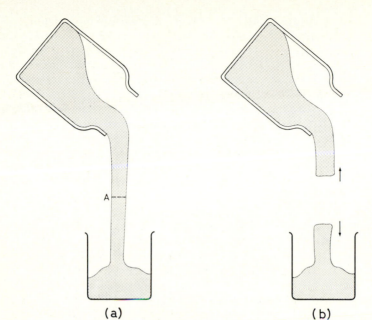

(a)   (b)

Fig. 11.3.   Cutting of elastic liquid, followed by retraction (adapted from
Lodge, 1964).

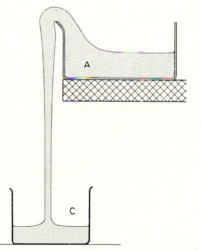

Fig. 11.4.   ' Self-siphoning ' property of poly(ethylene oxide).

of polymer processing.   In order to produce an object of any particular
form the material has to be shaped by the application of suitable forces;
this shape has then to be permanently fixed in one way or another.   In
rubbers, and in resins of the urea-formaldehyde type, this fixation of the

167

final form is brought about by a chemical cross-linking reaction. This, in effect, actually changes the material; once the cross-linking reaction has been carried out there is no further possibility of altering the form by any physical treatment. (This is why old tyres are so intractable; the rubber cannot be melted down and re-moulded. It can only be rendered usable by a *chemical* treatment which actually breaks down the molecules and destroys the network.) Most other polymers, for example, polythene, polyvinyl chloride, polystyrene, etc., are formed or shaped at a high temperature and subsequently set by cooling down to room temperature, the material being sufficiently rigid at room temperature to retain its shape indefinitely. These materials, which are softened by heat, are called *thermoplastic*; they may be re-heated and re-set any number of times.

The actual processes of formation encountered in industrial practice are many and varied. Sheets and films are usually produced by extrusion (in much the same way as in the production of filaments by the spinning operation) and may be finished by passing through a set of rolls, or calender*. Rods and tubes of all kinds are also produced by an extrusion process. Articles such as combs, buttons, engineering components, etc., are usually formed by injection moulding; in this process the molten polymer is forced, under high pressure, through a small aperture into a mould, where it is immediately cooled and solidified. Moulded articles are also produced from sheet or other shapes by direct pressure between plates. Bottles, beakers and hollow articles may be formed by blowing into a mould in much the same way as bottles are blown from glass. Another important process is spreading, as used in the application of a polymer to a base, e.g. in the production of coated paper, rubberized cloth, etc.

In these various processes the patterns of flow and the rates of flow are different; each of them involves at every stage of the operation a delicate relation between the temperature, the applied stress, and the viscosity and other properties of the polymer. For the satisfactory operation and control of such processes it is desirable to understand as much as possible about the fundamental flow properties of polymers and the way in which these properties may be measured and controlled.

Shear flow

The simplest type of flow to consider is shear flow. Let us imagine the liquid to be confined between two plates, the lower plate AB being fixed while the upper plate CD moves with constant velocity v as shown in fig. 11.5 (*a*). Under these conditions the liquid is being *sheared* at a uniform rate, g, given by

$$g = v/y \tag{11.2}$$

In order to maintain this state of flow it is necessary to apply equal and

* The word *calender* (from the Greek *kulindros*, roller) is allied to *cylinder*.

168

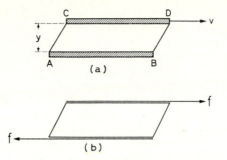

Fig. 11.5. (*a*) Simple shear flow. (*b*) Corresponding shear stress.

opposite forces to the top and bottom plates, i.e. to the top and bottom surfaces of the liquid (fig. 11.5 (*b*)). The forces act in the direction of shearing, and are tangential to the surfaces on which they act. The *shear stress* S is defined as the tangential force per unit area of the surface on which the force acts.

In simple liquids, such as water, the rate of shear is proportional to the shear stress (fig. 11.6 (*a*)). We may therefore write

$$S = \eta g, \tag{11.3}$$

where S is the shear stress and η (Greek eta) is a constant, called the *viscosity* of the liquid. This relation is known as Newton's law, and liquids which behave in this way are called *Newtonian* fluids.

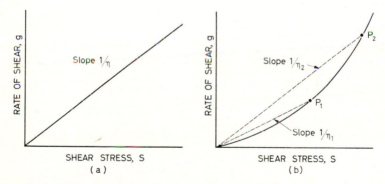

Fig. 11.6. Flow curves for (*a*) Newtonian and (*b*) Non-Newtonian fluids.

Molten polymers and polymer solutions do not, in general, behave in this way. A typical plot of rate of shear for such materials is shown in fig. 11.6 (*b*). The rate of shear is not proportional to the shear stress S, but increases more rapidly as S increases. The viscosity, which is defined as the ratio S/g, is no longer a constant, but *decreases* as the rate of shear is increased, as can be seen from the diagram. The slope of the

169

line OP_2 being greater than that of the line OP_1, it follows that $\eta_2 < \eta_1$. Only at very low shear rates is there a limited region in which the flow is approximately Newtonian.

This type of non-Newtonian flow has very important practical consequences. The reduction of viscosity with increasing rate of shear implies that the stresses required to produce high rates of flow are not as high as would be expected from measurements of viscosity at low shear rates. This applies, for example, to such operations as extrusion, in which the polymer is forced through a tube, to milling and mixing, which involve high shear rates, and also to rolling or calendering. A reduction of viscosity with increasing rate of shear is also an important property of paints (which are usually polymer solutions containing pigments). The application of a paint by brushing requires that the viscosity during this operation shall be not too high. But it is important that after application the paint will not run under the action of the very much smaller gravitational forces; this is ensured by the much higher viscosity of the liquid under this condition.

4. *The molecular mechanism of viscous flow*

In all liquids, including polymers, the viscosity is very sensitive to changes in temperature. For materials of low molecular weight, the dependence of viscosity on absolute temperature T is found experimentally to be given by the formula

$$\eta = A \exp (b/T) \qquad (11.4)$$

where A and b are constants. To understand this type of temperature dependence we must consider the molecular basis of viscous flow. The usually accepted theory of flow is that due to Eyring. Let us consider a liquid of low molecular-weight, in which the molecules are roughly spherical. The molecules in the liquid will pack together in a manner not very different from the state of packing in the solid or crystalline state, though somewhat less regularly. Eyring therefore represents the liquid structure as a regular structure, with, however, a certain number of holes or vacant sites. The accompanying fig. 11.7

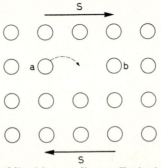

Fig. 11.7. Model of liquid according to Eyring's theory of viscosity.

shows such a structure containing a single vacant site. The molecules, which are subject to irregular thermal vibrations, are normally confined to fixed mean positions by the forces exerted by the surrounding molecules. Occasionally, however, a molecule next to a hole (e.g. the molecule a) will break away from its neighbours and 'jump' into the vacant space, leaving a hole in the position which it previously occupied. There is thus a constant change of structure arising from the jumping of molecules into holes. In the absence of an applied stress, these changes take place entirely at random: there is no preferred direction of jumping and hence no overall motion of the liquid. A shear stress, however, will introduce a directional effect. In the diagram, the stress S is tending to slide the top layer of molecules over the second layer, the second layer over the third, and so on. This extra 'push' will favour the jumping of the molecule a into the neighbouring vacant site, but will oppose the jumping of b into the same site. This preferential direction of jumping, averaged over all the molecules, will produce a resultant bulk movement or flow in the direction of the applied stress.

On the basis of this model, the process of flow takes place not by the sliding of all the molecules in the layer at the same time, but in a stepwise manner, by the independent jumping of individual molecules from one equilibrium position to another. This individual jumping process requires that the molecule shall break away from the attractive forces which bind it to its neighbours. This requires a certain amount of energy, called the 'activation energy', and it is only those molecules which, at any instant, happen to possess the necessary energy, that are capable of executing a jump. The probability that a given molecule will possess the required energy increases rapidly with increasing temperature, and can be expressed by a well-known relation. Making use of this relation, Eyring was able to relate the frequency of jumping to the temperature. From this it is a comparatively simple matter to derive the viscosity. This is expressed by the formula

$$\eta = A \exp (E/RT) \qquad (11.5)$$

where E is the 'activation energy' (for 1 mole) and R is the gas constant per mole.

Comparison of equation (11.5) with the experimental relation, equation (11.4), shows that the temperature coefficient of viscosity, as represented by the constant b in equation (11.4), is a measure of the 'jump' energy E. The value of this quantity, as derived from experiment, therefore tells us something about the magnitude of the forces which hold the molecules in their normal equilibrium positions in the liquid state.

5. *Application to polymers*

The application of Eyring's theory to the problem of flow in polymers leads to an interesting and rather unexpected result. The polymer

171

molecule is of course not spherical but very long. We should expect the energy required for it to break away from its surroundings to be proportional to the length of the chains, and hence to be very much greater for a polymer than for a small molecule containing the same kinds of atoms. Experiments on the temperature coefficients of viscosity for polymers show that these also obey the Eyring equation, but the values of the activation energy E come out not very much higher than in the case of low molecular-weight liquids. This can only mean that the ultimate unit in flow—the particle which jumps—has dimensions of the same order of magnitude in the two systems.

Following up this reasoning, Eyring studied a series of simple hydrocarbons of the paraffin type with various chain lengths. In fig. 11.8 the

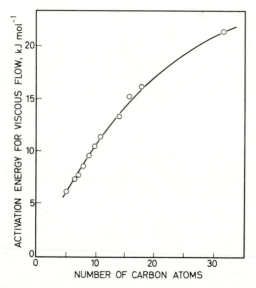

Fig. 11.8. Dependence of activation energy for viscous flow of paraffins on number of carbon atoms in chain (Kauzmann and Eyring, 1940).

resulting values of the ' activation energy ' are plotted against the number of carbon atoms in the chain. It is seen that this at first increases almost in proportion to the number of C atoms, but that with further increase in chain length the rate of increase of the activation energy becomes slower. Ultimately, this energy appears to be approaching a limit for a chain length of about thirty C atoms. Even for polythene (which may be regarded as a paraffin of infinite molecular weight) the value of E does not rise significantly higher.

These results show that in the case of a long-chain compound the unit which moves is not the whole molecule but a short segment of the chain, of length not greater than about thirty carbon atoms.

On reflection, this conclusion is perhaps less surprising than it might

172

appear to be at first sight. We have already seen that the polymer molecule acts like a chain of randomly jointed links, a chain, that is, in which the orientation of any one link bears no relation to that of neighbouring links. It is a natural consequence of this concept that the chain cannot move as a single rigid entity, but that individual short segments must move more or less independently of one another. The conclusion derived from the Eyring theory of viscosity is therefore not inconsistent with what is known about the form of the polymer molecule and the fluctuations of form brought about by internal rotations.

The way in which the actual detailed motion of a segment takes place will naturally vary according to the local geometrical disposition of the chain. A simple example, which will serve to illustrate the principle, is shown in fig. 11.9. In this, the short segment, consisting of the bonds 2, 3 and 4, is able to rotate as a unit about an axis formed by the bonds 1 and 5.

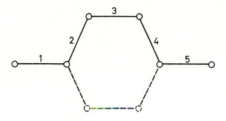

Fig. 11.9.   Illustration of particularly simple type of segmental motion due to rotation about bonds.

6. *Chain length and viscosity*

The fact that in a polymer the temperature dependence of the viscosity, and the activation energy derived from it, are both independent of the chain length does *not* mean that the actual value of the viscosity is similarly independent. This is far from being the case. The value of the viscosity is determined by the resultant motion of the polymer molecule *as a whole*. This resultant motion depends not only on the rate of ' jumping ' of individual segments, but also on the way in which these segmental jumps are related to one another in direction. Because of the complicated form of the molecule in space, and the mutual entanglements between molecules (discussed in Chapter 3) not all individual jumps are equally effective in producing a resultant displacement of the molecule. Consider, for example, a situation such as that represented in fig. 11.10, in which a molecule ABC is interlooped with another molecule at D, this molecule itself being temporarily held in position by entanglements at E and F. Let us suppose that the shear stress f produces displacement of the segment BC to the position BC'. This, in itself, will cause a displacement of the centre of mass of the molecule to the right. The effect of this displacement will be to introduce a tension into the chain segment

173

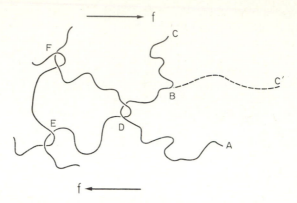

Fig. 11.10.　Illustration of possible changes in chain conformation during flow.

DBC (see Chapter 3), which will tend to pull the remainder of the molecule (DA) through the loop.　The resultant motion of this segment DA will be in the opposite direction to the original displacement of the segment BC, as a result of which the centre of mass of the molecule will be displaced to the left.　This second displacement therefore *reduces* the contribution of the molecule to the overall flow.

The relation between viscosity and molecular weight for a typical polymer (polyisobutylene) as determined by experiment, is shown in fig. 11.11.　Plotted on a logarithmic scale, a sharp discontinuity is found at a particular value of molecular weight.　Above this point the slope of the line is 3·4, which means that the viscosity is proportional to the 3·4 power of molecular weight.　Below the discontinuity the slope is much lower, namely about 1·75.　A similar discontinuity is found with other polymers, but the value of the molecular weight at which it occurs varies from one polymer to another.

It is believed that the discontinuity represents the point at which the effect of entanglements between chains becomes predominant.　Below this point the molecules are considered to move more or less individually, as in an ordinary liquid; above this point any given molecule becomes so entangled that it can only move by dragging other molecules along with it.　Since these entanglements increase greatly in complexity as the chain length increases, this process leads to a very high rate of increase of viscosity with increasing molecular weight.

7. *Weissenberg effects*

In the preceding discussion of the nature of shear flow it has been tacitly assumed that the only stress present in the flowing liquid is a shear stress acting *tangentially* to the surfaces in the direction of sliding or shearing (fig. 11.5).　This is a correct assumption for ordinary liquids,

174

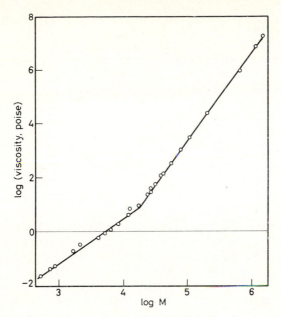

Fig. 11.11.   Dependence of viscosity of polyisobutylene on molecular
weight M (Fox and Flory, 1951).

but it is not generally true for polymers at high rates of shearing.   In
these materials, the temporary molecular network which exists becomes
elastically deformed during flow, and this elastic deformation generates
stresses of a different kind within the flowing liquid, in addition to the
ordinary shearing stress.   These stresses may be regarded as equivalent
to an internal tension, or tensile stress, acting in a direction parallel to the
direction of sliding on planes perpendicular to this direction (fig. 11.12).

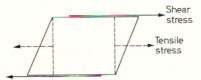

Fig. 11.12.   Tensile stress developed in elastic liquid during shear flow.

Stresses of this kind, which are perpendicular or *normal* to the planes on
which they act, lead to some very curious effects, called ' Weissenberg
effects ', after Weissenberg, who made a special study of these phenomena, and designed instruments for the measurement of both the normal
and the shear stresses present during flow.

   Consider a coaxial cylinder system such as that shown in fig. 11.13.
The liquid in the gap between the inner and outer cylinders is subject to

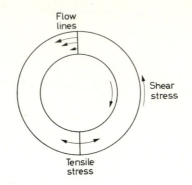

Fig. 11.13. Normal stress in coaxial-cylinder system.

a shearing action, which generates a shear stress on the inner and outer cylinders, and *in addition*, a tensile stress in the direction of the stream-lines. This tensile stress cannot be directly observed, but, owing to the fact that it is acting round the circumference of a circle, it gives rise to an internal *pressure* in the liquid, just as if the liquid were surrounded by a number of stretched elastic bands. This pressure forces the liquid up the walls of the inner cylinder.

It makes no difference whether the inner cylinder is fixed while the outer is rotated or *vice versa*. The effect is most readily demonstrated by rotating the inner cylinder in a fixed outer cylinder. For the purpose of demonstration, the inner cylinder may conveniently be a rod of about 1 cm diameter, while the outer ' cylinder ' may be a beaker of diameter, say 5–10 cm. The rod may conveniently be rotated by a hand drill fixed in a suitable clamp (fig. 11.14). On rotation, the liquid ' climbs ' up the rod to a height of several cm*.

Weissenberg effects are of undoubted interest in connection with any polymer process, such as extrusion, which involves continuous flow, for they give rise to dimensional changes in the material. A good example is the melt-spinning process used for the production of filaments from molten polymers. In this process the liquid emerging from the spin-neret (fig. 8.4) is seen to expand in diameter to several times the diameter of the spinneret aperture, giving the onion-shaped profile shown on an enlarged scale in fig. 11.15. This extrusion swelling, known in current jargon as ' die swell ', is probably due to a combination of normal stress effects and of the ordinary elastic recovery resulting from the previous compression of the liquid as it enters the die. The fact that it is not wholly due to the latter cause is shown by experiments involving flow through tubes of different length. The amount of expansion is found to

* A very suitable demonstration liquid is ' TRAPT ' rat killer, a rubbery composition obtainable from William Charles Knox, Esq., 6 Ferndale St., Belfast, N. Ireland. This is not poisonous: the rats are unable to escape from the very sticky medium which entraps them.

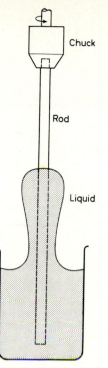

Fig. 11.14.   Demonstration of ' rod-climbing ' effect in coaxial cylinder system.

Fig. 11.15.   Expansion of liquid thread on emerging from spinneret.

be a function of the length of the tube up to a certain length, beyond which it becomes constant. In this constant region the tube length is so great that the conditions of flow in the exit region are no longer affected by the stresses to which the polymer was subjected before entry into the tube. Under these conditions the swelling must be a genuine consequence of the steady flow in the tube.

The expansion on extrusion may also be observed in polymer solutions. A convenient system for demonstration is a 4% solution of perspex in dimethyl phthalate contained in a wide tube whose end is drawn down to a nozzle of diameter 1 to 2 mm. The liquid is forced through the tube by the application of a suitable air pressure (e.g. from a bicycle pump). A lateral expansion to two to three times the diameter of the nozzle is readily obtained.

8. *Conclusion*

The examples chosen to illustrate some of the characteristic flow properties of polymers will serve to indicate their remarkable range and variety. In no other aspect of mechanical properties are polymers so distinctively different in their behaviour from ordinary low molecular-weight materials. The unique combination of a viscosity which depends on the rate of shearing with large elastic deformations of the type encountered with rubbers produces effects which are different not only in amount, but also in kind, from any observed with non-polymeric materials. While the general basis of these effects is capable of a reasonably satisfactory molecular explanation, a fully worked-out theory capable of yielding a precise quantitative formulation of many of the observed effects does not at present exist. The problems discussed in this chapter, however, are of great theoretical as well as practical and industrial interest, and there seems little doubt that in the course of the next few years more effective theoretical methods of dealing with them will emerge.

References

BEKKEDAHL. *Bur. Stand. J. Res.*, **13**, 410, 1934.
BEKKEDAHL and WOOD. *Ind. Engng Chem.*, **33**, 381, 1941.
BERRY. *J. Polym. Sci.*, **50**, 313, 1961.
BUNN. *Trans. Faraday Soc.*, **35**, 482, 1939.
CURTIS and TRELOAR. *Nature*, **220**, 60, 1968.
DART, ANTHONY and GUTH. *Ind. Engng Chem.*, **34**, 1340, 1942.
FLORY. *Ind. Engng Chem.*, **38**, 417, 1946.
FOX and FLORY. *J. Phys. Chem.*, **55**, 221, 1951.
GORDON, MANFRED. *High Polymers* (Iliffe), p. 41, 1957.
GORDON and MACNAB. *Trans. Faraday Soc.*, **49**, 31, 1953.
GUTHRIE. *J. Text. Inst.*, **40**, T489, 1949.
HALL. *J. appl. Polym. Sci.*, **8**, 237, 1964.
HOLLAND and TURNER. *J. Soc. Glass Technol. Trans.*, **18**, 225, 1934.
JAMES and GUTH. *J. chem. Phys.*, **11**, 455, 1943.
JEFFRIES. *J. Text. Inst.*, **51**, T339, 399 and 441, 1960.
JOULE. *Philos. Trans.*, **149**, 91, 1859.
KAUZMANN and EYRING. *J. Am. chem. Soc.*, **62**, 3113, 1940.
KELLER, LESTER and MORGAN. *Philos. Trans. A*, **247**, 1, 1954.
LODGE. *Elastic Liquids* (Academic Press), p. 238, 1964.
MANDELKERN. *Crystallization of Polymers* (McGraw-Hill), pp. 106, 218, 1964.
MEREDITH. *Mechanical Properties of Textile Fibres* (North-Holland), p. 73, 1956.
MEYER and FERRI. *Helv. chim. Acta*, **18**, 570, 1935.
MOORE and WATSON. *J. Polym. Sci.*, **19**, 237, 1956.
MULLINS. *I.R.I. Trans.*, **22**, 235, 1947.
WOOD. *Advances in Colloid Science*, II (Interscience), p. 57, 1946.

INDEX

(Figures in italic refer to chapters)

182

THE WYKEHAM SCIENCE SERIES
for schools and universities

1 *Elementary Science of Metals*
(S.B. No. 85109 010 9)

J. W. MARTIN and R. A. HULL
20s.—£1.00 net *in U.K. only*

2 *Neutron Physics*
(S.B. No. 85109 020 6)

G. E. BACON and G. R. NOAKES
20s.—£1.00 net *in U.K. only*

3 *Essentials of Meteorology*
(S.B. No. 85109 040 0)

D. H. MCINTOSH,
A. S. THOM and V. T. SAUNDERS
20s.—£1.00 net *in U.K. only*

4 *Nuclear Fusion*
(S.B. No. 85109 050 8)

H. R. HULME and A. McB. COLLIEU
20s.—£1.00 net *in U.K. only*

5 *Water Waves*
(S.B. No. 85109 060 5)

N. F. BARBER and G. GHEY
20s.—£1.00 net *in U.K. only*

6 *Gravity and the Earth*
(S.B. No. 85109 070 2)

A. H. COOK and V. T. SAUNDERS
20s.—£1.00 net *in U.K. only*

7 *Relativity and High Energy Physics*
(S.B. No. 85109 080 X)

W. G. V. ROSSER
and R. K. MCCULLOCH
20s.—£1.00 net *in U.K. only*

8 *The Method of Science*
(ISBN 0 85109 090 7)

R. HARRÉ and D. EASTWOOD
25s.—£1.25 net *in U.K. only*

9 *Introduction to Polymer Science*
(ISBN 0 85109 100 8)

L. R. G. TRELOAR
and W. F. ARCHENHOLD
30s.—£1.5 net *in U.K. only*

10 *The Stars: their structure and evolution*
(ISBN 0 85109 110 5)

R. J. TAYLER
and A. S. EVEREST
30s.—£1.5 net *in U.K. only*

11 *Superconductivity*
(ISBN 0 85109 120 2)

A. W. D. TAYLOR and G. R. NOAKES
25s.—£1.25 net *in U.K. only*

THE WYKEHAM TECHNOLOGICAL SERIES
for universities and institutes of technology

1 *Frequency Conversion*
(S.B. No. 85109 030 3)

J. THOMSON,
W. E. TURK and M. BEESLEY

2 *The Art and Science of Electrical Measuring Instruments*
(ISBN 0 85109 130 X)

E. HANDSCOMBE

Price per book for the Technological Series 25s.—£1.25 net *in U.K. only*